Effective
Maintenance

Effective Maintenance

THE KEY TO PROFITABILITY

A Manager's Guide to Effective Industrial Maintenance Management

Paul D. Tomlingson

VNR VAN NOSTRAND REINHOLD COMPANY
New York

Library of Congress Catalog Card Number 93–70
ISBN 0–442–00436–2

I(T)P   Van Nostrand Reinhold is a division of International Thomson Publishing.
         ITP logo is a trademark under license.

Printed in the United States of America

Van Nostrand Reinhold
115 Fifth Avenue
New York, NY 10003

International Thomson Publishing
Berkshire House
168–173 High Holborn
London WC1V7AA, England

Thomas Nelson Australia
102 Dodds Street
South Melbourne 3205
Victoria, Australia

Nelson Canada
1120 Birchmount Road
Scarborough, Ontario
M1K 5G4, Canada

16 15 14 13 12 11 10 9 8 7 6 5 4 3 2 1

Library of Congress Cataloging-in-Publication Data

Tomlingson, Paul D.
    Effective maintenance : The key to profitability : a managers
guide to effective industrial maintenance management / Paul D.
Tomlingson. — 1st ed.
        p.   cm.
    Includes index.
    ISBN 0–442–00436–2
    1. Plant maintenance.   2. Plant maintenance—Management.
I. Title.
TS192.T66   1992
658.2'02–dc20   93–70
                                                           CIP

Contents

Preface

Effective Maintenance: The Key to Profitability is divided into three sections.

Part I How should maintenance organize, define, and execute its program to best support your operations?

Part II What should you examine to establish that: The plant enviroment is conducive to the success of maintenance? Maintenance is organized to respond quickly and effectively to its objectives while making the best use of its personnel? The maintenance program is well-defined and understood by those who carry it out, use its services (operations), or support it (staff departments)? Maintenance performance is assessed and steps are taken to achieve improvement?

Part III What evaluation and improvement strategy will ensure successful maintenance performance?

Against these questions, the book examines the maintenance organization, its programs and controls, how other departments must cooperate and support maintenance, and management actions to ensure the success of maintenance. Also, the book suggests a management strategy for ensuring the improvement of maintenance and its continuing, effective contribution to the profitability of your plant.

Introduction

This book has the objective of helping plant and facility managers to gain better performance from their maintenance organizations and ensure profitability. It will help you to:

1. Establish what maintenance should be doing in your plant enviroment.
2. Determine whether maintenance is organized correctly.
3. Find out whether maintenance is performing effectively.
4. Establish an improvement program if needed.
5. Ensure continuous improvement and effective performance.

However, before you can assess maintenance or devise corrective actions, you must take several fundamental actions to create an environment in which your maintenance organization and its program can be successful. While maintenance provides a service to production, operational circumstances, like meeting higher production targets, could allow production to defer essential maintenance services. Similarly, maintenance depends, for example, on the warehouse for repair materials and on accounting for information to make repair decisions. Thus, maintenance typically controls little of its own destiny. Therefore, it requires your help.

You can help them to succeed by:

1. Ensuring that maintenance has a clear objective: A statement of maintenance tasks, their priorities, and limitations in supporting your plant's overall production strategy (your plan for achieving profitability).
2. Providing policies guidelines: Specific roles and responsibilities that ensure harmonious interaction between departments and, incidentally, help ensure that the maintenance program can be successful.

With effective guidelines you can:

1. Provide maintenance with the guidance they need to organize themselves, define their program, and control the work they perform.
2. Guide production in actions to utilize maintenance services effectively.
3. Guide the warehouse, purchasing, accounting, or data processing in supporting the maintenance function.

Thus, your clarification of objectives and provision of policy guidelines are vital actions in creating the enviroment for the success of maintenance.

About the Author

Paul D. Tomlingson is a management consultant specializing in the design, implementation, and evaluation of maintenance management programs for industry. He is a veteran of 22 years of world-wide maintenance management consulting, the author of three textbooks and over 85 published trade journal articles on maintenance management. Tomlingson is a 1953 graduate of West Point and in addition to a B.S. in Engineering, he holds an M.A. in Government and an M.B.A., both from the University of New Hampshire. In addition to the presentation of his own public and on-site seminars, he has appeared in seminars at the University of Wisconsin, Hofstra University, the University of Denver, National and Regional Plant Engineering Shows, and the Society of Mining Engineers Conferences. Mr. Tomlingson has been listed in *Who's Who in the West*.

I

How Should Maintenance Organize, Define, and Execute Its Program to Best Support Your Operations?

1

Maintenance Organization

How should maintenance organize itself to best support your plant maintenance requirements?

MAINTENANCE TASKS

Principal Tasks

The industrial maintenance organization must be able to perform a multitude of tasks effectively, including the following:

Support operations by keeping production equipment in good condition so that production targets can be met.

Maintain the plant by keeping the plant site and its buildings, utilities, and grounds in a functional, attractive condition.

Conduct engineering projects like construction, equipment modification, and installation or relocation.

Develop a program for carrying out its services.

Organize itself to support the equipment maintenance needs of production while conducting essential engineering projects.

Execute its program while utilizing its resources productively.

Perform quality work.

Anticipate and prepare for future work.

Achieve continued improvement by evaluating performance, taking corrective actions, and measuring progress.

Prepare for future changes by anticipating needs and organizing flexibly.

VERIFY ORGANIZATIONAL PRINCIPLES

Applying Organizational Principles

In organizing themselves, maintenance must successfully apply organizational principles. They must typically adhere to the following:

Assign specific objectives to units to promote understanding and mutual support.

Keep different units of the organization compatible so that they work harmoniously.

Specify the functions to be performed.

Allow the organization to achieve efficiencies without imposing restraints.

Set common goals so that all supervisors or teams know what is expected of them.

Ensure that the maintenance task is divided so that each supervisor or team has an amount of work he/she or they can successfully handle.

Assign responsibility for the whole function to one individual or a single team.

Ensure sufficient authority is provided to carry out assigned responsibilities.

Assign only the number of personnel the supervisor can control or the team can utilize effectively.

Ensure all know exactly what duties they are responsible for.

Equalize responsibilities commensurate with the ability of personnel to carry them out.

Avoid inhibiting capabilities with unnecessary constraints or rules.

If something takes care of itself, leave it alone. Alternately, report problems rather than unnecessary confirmation of normal situations.

TROUBLE SIGNS

What to Look For

The danger signs of poor maintenance organization often reveal that:

By not identifying its real objective (maintain production equipment), maintenance may accept tasks thrust on them (equipment modification) and react rather than organize.

The work that must be done to achieve the correct maintenance objectives is rarely identified properly. Typically, without timely equipment inspections, surprises force emergency repairs and few problems are uncovered soon enough to plan the work.

Often the only division of work is to relate work to the jurisdictional areas that the labor contract permits. This results in the poorest use of personnel.

Assignment of specific work is overridden by craft jurisdiction, areas of supervisory familiarity, or tradition—none of which ensure worker productivity.

Maintenance tries too hard to promote from within the organization. A good mechanic does not always make a good supervisor. The best supervisors are those who control the work and effectively use the efforts of their crew members. Poorer supervisors try to work alongside crew members thus, diminishing their control. When these supervisors are given bigger responsibilities, they are often not equipped to manage them.

Maintenance can be a thankless job. As a result, personnel outside of maintenance are not always anxious to join the maintenance organization. Thus, qualified personnel are excluded in favor of those within maintenance who may be less qualified.

If maintenance procedures are not based on policies, few work effectively.

DUTIES OF KEY PERSONNEL

Verify What Key Personnel Do

There is a tendency for key personnel to view their jobs one level, or sometimes two levels, below what they should be doing.

Superintendents supervise work instead of planning ahead and monitoring cost and performance.

General supervisors are often more technically-oriented and uncomfortable with the "people-skills" they should be able to use fluently.

Supervisors may emphasize their favored skills by doing work rather than by supervising their crews.

See Appendix A, Outline of Duties of Key Maintenance Personnel.

ORGANIZATION

Types of Organizations

The three most prominent and useful types of industrial maintenance organizations include the craft organization, the area organization, and the team organization. While each of these types has distinct characteristics, most are "hybrids."

Craft Organization

In the craft organization, one maintenance supervisor controls a crew made up entirely of the same craft—millwrights, mechanics, electricians, and so forth. See Figure 1–1.

The craft organization is being phased out gradually as managers realize that it is less flexible, often yields poor productivity, and is linked too often with labor contracts emphasizing inflexible craft jurisdictional boundaries.

Area Organization

The area organization provides a geographical orientation to maintenance. The area organization makes a single maintenance supervisor responsible for all maintenance carried out in a well-identified, reasonably-sized geographical area. The area maintenance supervisor usually has a multicraft crew, made up of the crafts necessary to perform all types of field work in his/her area. See Figure 1–2.

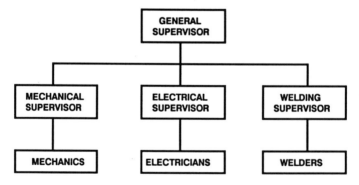

FIGURE 1–1.   One maintenance supervisor controls a single craft in the craft organization.

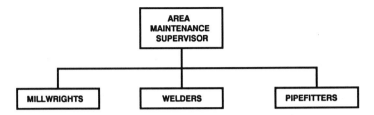

FIGURE 1–2. The area maintenance supervisor has all of the crafts necessary to perform day to day maintenance.

The area crew performs day to day maintenance (PM inspections, emergency repairs, and unscheduled work). They might obtain labor reinforcements from a centrally controlled craft organization to meet peak workloads like a shutdown. Rarely, is there a pure area organization. More often, the workload dictates that no one area can use all crafts needed on a full-time basis. For example, an area crew might need 8 full-time millwrights but, could use only 12 hours per week of a pipefitter. Therefore, these crafts would be centrally controlled and loaned to the area only when needed.

Craft Pool

The use of a centrally controlled craft pool in conjunction with the area organization can improve labor utilization. The craft pool can minimize the size of the area crews because it offers the capability of rapidly reinforcing the area crews to meet peak workloads. Once the peak workload has been met, craft pool members are withdrawn and assigned to other areas with peak work loads. See Figure 1–3.

The Team Concept

Many plants have adopted teams in which a team leader who, rather than give definitive orders, looks for consensus among other team members. Once agreement is reached, they act as a group. In other instances, the team leader heads up both maintenance and operations. He/she relies on team members to take individual and group actions based on a consensus that they reach within general guidelines provided by him/her. See Figure 1–4.

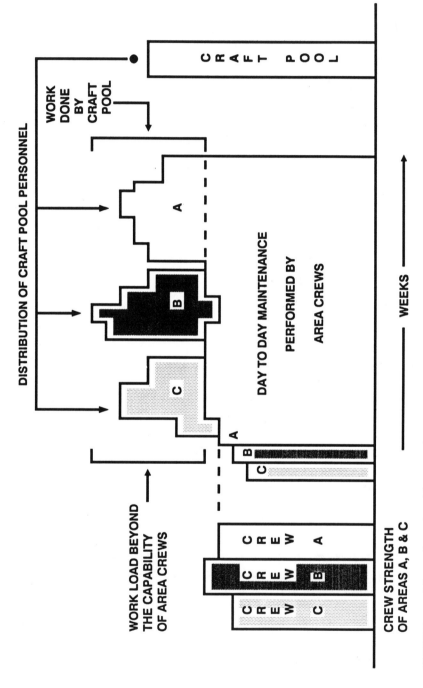

FIGURE 1-3.   The craft pool is able to reinforce the area organization to meet peak work loads.

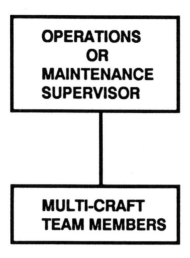

FIGURE 1–4. The team organization pools the talent of several different crafts and provides opportunity for team decision making on work control matters.

In some cases, the team members are salaried. Plants using the team concept are usually unencumbered by unions and are relatively free to organize as local management sees fit. This can yield significant productivity improvements. For example:

A paper mill using the team concept has a 127-person maintenance workforce compared to 319 for a competing union mill with the same product output.

A popular aspect of the team is the use of operators in performing some maintenance. Good results are being achieved by operators in performing lubrication, minor adjustments and small component replacements (belts, guards, filters, and so on), providing that they are trained.

If you are considering the team organization, ensure that the team is developed within the framework of a well-defined maintenance program.

Similarly, if you are converting from a traditional organization to a team, ensure that the supervisors are assigned a specific role that both they and you consider essential.

CONTRACT MAINTENANCE

Primary or Supplementary

Many plants use contract maintenance for all maintenance or to supplement the regular maintenance workforce during peak periods such as a shutdown. One of the greatest appeals of contract maintenance is its ability to staff up when needed and reduce its size when not needed. However, contract maintenance will be expensive so it must be used wisely. The most effective contract maintenance appears where multiple plants find it convenient to shift contract personnel quickly between locations. The company manages the maintenance function with its own program. It then controls the contract workforce through the company's own key personnel from superintendent down to general supervisor, planners, and maintenance engineers. The contractor provides field supervisors and craftsmen to execute the work. Beware of the contractor who shows up with everything, including his/her own maintenance program. Within a short time the program is so imbedded in the plant's accounting procedures that the plant cannot get rid of the contractor even when they know they must.

ASSIGNING MAINTENANCE WITHIN THE PLANT ORGANIZATION

Control of Maintenance

Maintenance departments can be controlled by plant or production managers. Control can also be fragmented. For example, plant services might be controlled by the plant engineer and production maintenance controlled by the production superintendent. There is no rule of thumb nor is there any best solution. Like so many other organizational solutions, maintenance is often built around the talent that is considered able to control it best. However, if the organization will be headed by a non-maintenance leader, ensure that the program is well-defined.

Summary

The objective that you assign to maintenance should dictate what the maintenance organization must do. Will their focus be entirely on the maintenance of production equipment? Will you allow them to support engineering projects, and, if so, to what degree? You should check how well they have adhered to the principles of organization. Review the traditional organizational steps. Trace the responsibilities of key personnel to ensure that they both support the maintenance objective but also focus on the right activities. Examine the type of organization to determine whether or not it can support the assigned objective properly. Do not overlook the possibility of a team, but make sure the maintenance program is well-defined before you introduce the team concept. Consider the use of a contractor, but be careful how their activities will be controlled. As the new organization settles in, be aware that it will require changes, some sooner than anticipated. Therefore, appreciate the need for future organizational change.

2

Maintenance Program

Is the maintenance program defined so that maintenance personnel can perform it effectively, operations can use its services properly, and staff personnel will understand how to support it conscientiously?

THE PRODUCTION STRATEGY

Maintenance Program Versus Production Strategy

The maintenance program is an integral part of the plant production strategy—the overall plan by which plant departments work harmoniously toward a common objective of profitability. Within your strategy, each principal department should have a clear objective. All objectives should be mutually supporting and contribute directly to the common plant goal of profitability. Accompanying the objective are your policy guidelines which amplify its intent and spell out responsibilities to ensure the joint, cooperative efforts of plant departments.

DEFINING MAINTENANCE

What Maintenance Must Do

Guided by your assigned objective and policy guidelines, maintenance should establish day-to-day procedures and fit them into a framework depicting how maintenance services are requested, planned, scheduled, controlled, and measured. They should define terminology used to ensure that it is understood. To adequately define the maintenance program, you should provide:

An Objective: A clear statement of what maintenance is to do in supporting the plant's overall production strategy.

Policy Guidelines: Identification of specific roles and responsibilities which must be carried out by each department to ensure the maintenance program can be successful.

Maintenance then must:

Develop a Concept: An overview of the maintenance program detailing the specific manner in which work is requested, planned, scheduled, executed, controlled, and measured.

Establish Procedures: Details to allow personnel to easily follow instructions for obtaining or carrying out maintenance services.

Define Terminology: Establish common maintenance terms to promote understanding.

Collectively, these elements constitute maintenance program definition. Once the program is defined an education program must follow to ensure all appropriate plant personnel can perform it effectively, use its services properly, and support it conscientiously.

The following appendices illustrate a maintenance objective, typical policies, workload definition, and common maintenance terminology:

Appendix B   Typical Maintenance Objective
Appendix C   Illustrative Policy Guidelines for Maintenance
Appendix D   Maintenance Workload Definition
Appendix E   Maintenance Terminology

THE CONCEPT OF MAINTENANCE

Program Definition Avoids Confusion

Imagine what might happen in your plant if maintenance neglected to provide answers to the following questions:

1. How are maintenance services requested?
2. Once received by maintenance, how do they go about responding?
3. What keeps production from calling everything an emergency?
4. What kind of work does maintenance plan?
5. What criteria exist to distinguish non-maintenance from maintenance work?
6. Who reports labor use?
7. Who attends the scheduling meeting?
8. How is work status reported?
9. What is the maintenance engineer supposed to do?
10. Who approves major work like overhauls?

These questions must be answered. If they are not, confusion, a lack of control, and poor maintenance result. Maintenance organizations have tried to answer these questions in many ways. Some have written standing operating procedures. Others have developed massive procedure manuals. While all of these are helpful, they often only permit maintenance to say, "Yes, we have documented our program." Unfortunately, such documentation is seldom read. Less often is it the subject of an educational effort. The result is little clarity as to how the program operates, what roles personnel play, how control is achieved, and how performance is measured. An increasingly necessary technique to adequately convey the framework of how maintenance services are requested, prepared for, performed, and controlled is the concept of maintenance. See Figure 2–1.

Summary

The benefit of a quality maintenance program to a plant's profitability is lost if that program, for whatever reason, is not fully effective. Many potentially effective maintenance programs are rendered ineffective only because they are forced to operate in an environment where the program can be ignored with devastating consequences. To avoid this you must establish a clear objective for maintenance and provide policies prescribing how maintenance will be performed. Then, based on your guidance, maintenance must define its program so that all appropriate plant personnel can perform it effectively, use its services properly, and support it conscientiously.

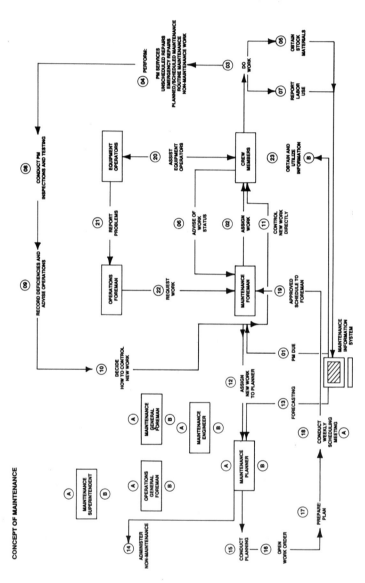

CONCEPT OF MAINTENANCE

FIGURE 2–1. Concept of Maintenance. The Concept of Maintenance illustrates a framework depicting how maintenance services are requested and thereafter work planned, scheduled, executed, controlled, and measured to determine the effectiveness with which work was carried out. The concept also shows how key maintenance personnel interact while using system elements to carry out the program. Concurrently, it depicts the actions of key operating, staff, and management personnel as they utilize maintenance services or support the maintenance program.

16

1. The computer provides the maintenance supervisor with information on preventive maintenance services due. They, in turn, make assignments to their crew members. Alternately, the planner might advise the supervisor of services due.

2. The maintenance foreman makes work assignments to the crew members using the work order system. Since verbal orders are often used, they must be a legitimate part of the work order system.

3. Crew members receive work assignments and prepare to do work.

4. Crew members perform the following: preventive maintenance services, unscheduled repairs, emergency repairs, planned and scheduled maintenance, routine maintenance, and non-maintenance work such as construction.

5. In preparation for doing the work, or while work is in progress, crew members obtain stock materials. Completed stock issue cards flow into the data base to adjust inventory levels and provide information to maintenance on stock material costs and usage.

6. As they do work, or at its completion, crew members advise their supervisor of the status of the work assigned to them.

7. Upon work completion, crew members use the time card to report man-hours spent on each job. After verification by the foreman, time card data flows into the information system where man-hours are converted into labor costs.

8. PM inspections and nondestructive testing (like vibration analysis) are used to monitor equipment condition and help uncover equipment deficiencies.

9. When equipment deficiencies are found they are classified as work requiring planning, unscheduled or emergency repairs. It is advisable to inform the operations foreman of deficiencies found during a PM inspection so that he:
   a. Is aware of the inspection effort.
   b. May anticipate the need for repairs.
   c. Knows maintenance will follow through on the work.

10. As the maintenance foreman is advised of work requirements he may:
   a. Make direct assignments to crew members (verbal orders).
   b. Forward jobs to the planner if they meet the planning criteria.
   c. Hold or store jobs in the computer to be done later.

11. Unscheduled and emergency repairs are controlled directly by the maintenance foreman since they require no formal planning.

(continued)

FIGURE 2-1. Continued.

12. Work requiring planning is identified and sent to the planner.

13. Using the forecast, the maintenance planner determines the periodic maintenance actions that are due over the next several weeks. He verifies, through the latest PM inspections, that the components do, in fact, warrant replacement. Since most forecasted jobs will have been done many times in the past, they will have a standardized tasklist (what to do) as well as a standard list of required materials and tools. He also allows enough time to assemble the required materials. The planner advises the warehouse supervisor or purchasing agent of materials needed. Maximum lead time should be allowed for assembling the materials. As materials for specific planned jobs become available, the planner is advised.

14. Some work to be performed by maintenance will be non-maintenance projects such as construction, equipment installation, and so forth. This type of work should be classified to distinguish it from maintenance work. Also, it should be evaluated to ensure that it is necessary, feasible, and properly funded. It should be prioritized because it will compete for resources that are also required for maintenance work. If a significant amount of non-maintenance work is required, management should establish limits to avoid overcommitment of maintenance resources.

15. As each job is planned, the planner would do the following:

    a. Establish the job scope.
    b. Conduct a field investigation or use an existing standard to develop a job plan.
    c. Estimate manpower by craft and sequence into the job plan.
    d. Identify materials, tools, rigging, and support equipment needs.
    e. Estimate job cost.
    f. Establish the future week during which the work should be performed.
    g. Establish job priority with the equipment user (operations).
    h. Obtain job approval.
    i. As required, the maintenance engineer will help the planner to develop the job scope of any unique major jobs.

16. The work order is then opened. This authorizes the procurement of materials, the ordering of shop support, and the subsequent use of labor to perform the work (A).

17. For each job that is a candidate for inclusion in the following weeks schedule, the planner prepares a preliminary plan in which he verifies that, subject to operations approval, maintenance has the necessary resources to perform the work.

18

18. A scheduling meeting is conducted in which maintenance supervision presents the weekly plan to operations. Approval of the weekly plan by operations binds them to make the equipment available at the prescribed time and requires maintenance to make its resources available to do the work. At this point, the weekly plan becomes an approved schedule. During the scheduling meeting, compliance with the previous weeks schedule and performance on selected major jobs should be verified. Preparations for selected important jobs that are scheduled to be done in several weeks should be assessed also.

19. The approved schedule is presented to the maintenance foreman who will accomplish the work. The planner then coordinates with the foreman to arrange the timing of events such as on-site material delivery. Daily coordination meetings are held between operations and maintenance to adjust-ment the schedule in the event of operational delays.

20. Crew members assist equipment operators on problems as required. In some instances, equipment operators may report problems directly to workers as a result of their observations.

21. Operators request help from maintenance crew members informally for small, simple problems. However, major problems are reported to the operations foreman because greater resources and more downtime may be necessary to make repairs.

22. The operations foreman requests necessary work using the work order system.

23. The maintenance information system should provide information on the following:

a. Utilization of labor (including overtime use and absenteeism).

b. The status of cost and performance of selected planned jobs.

c. The backlog of pending jobs to help determine whether maintenance is keeping up with the generation of new work.

d. Repair history so that chronic, repetitive problems and failure trends may be uncovered and corrected and the life span of critical major components verified.

e. The cost of maintenance against units and components for the current month and year to date. In addition, operations should be advised of all jobs completed, unit by unit, each week. Maintenance personnel should be able to research part numbers, drawings, wiring or assembly diagrams using the computer. Management should be provided downtime data and performance indices like cost per ton.

3

Preventive Maintenance

What can you do to ensure effective preventive maintenance?
How should maintenance organize and conduct preventive maintenance services?
How can operations best support preventive maintenance?

MANAGEMENT SUPPORT

PM Policy

The most important step that you can take to ensure the success of preventive maintenance is to establish policies that prescribe how maintenance is to carry it out and how operations is to support it. In appropriate policy, maintenance will conduct a "detection oriented" preventive maintenance (PM) program. The program will include equipment inspection and nondestructive testing (predictive maintenance) to help avoid premature failure. Lubrication, servicing, cleaning, adjusting, and minor component replacement will be carried out to extend equipment life. PM will take precedence over every aspect of maintenance except bona fide emergency work. Production will perform PM-related tasks, such as cleaning and adjustment, and ensure that all PM services due are carried out on time. Compliance with the PM schedule will be measured and management will be informed of performance improvements as a result of PM services performed. The purpose of PM as well as the requirement of production to support the program and your interest in assessing compliance and benefits secured are demonstrated by such a policy.

OPERATIONS SUPPORT

Operations Approval

Once the preventive maintenance program has been organized, the operations superintendent should approve the PM program recommended to him/her by the maintenance superintendent. He/she then ensures that his/her key personnel understand the program and are aware of the supporting roles that they must play to ensure program success. Operations supervisors should monitor the overall effect of the PM program in reducing downtime. They should verify that scheduled PM services in their areas are carried out on time. Operations supervisors should require that maintenance inform them of deficiencies uncovered by inspections. Thereby, decisions can be made regarding the scheduling of equipment for repairs. Process engineers should verify that maintenance service checklists meet the technical needs of equipment operation and production processes. Operators should carry out tasks like cleaning and adjusting equipment or checking oil or hydraulic fluid levels. They can also help by operating equipment properly and reporting problems promptly. Mobile equipment operators should perform "before, during, and after operations" checks. Such checks are a valuable addition to the regular PM program. Since operators are on or near equipment continually, they can spot and report potential problems.

UNDERSTANDING PREVENTIVE MAINTENANCE

Defining Preventive Maintenance

Keep PM simple. It is equipment inspection and testing to avoid premature equipment failures, and lubrication, cleaning, adjusting, and minor component replacement to extend equipment life.

Purpose of Preventive Maintenance

PM services help to avoid equipment failures through the use of:

Equipment inspections to uncover deficiencies before failure and in sufficient time, plan deliberate repairs.

Nondestructive testing techniques (predictive maintenance) to detect equipment deterioration and monitor equipment condition to note abnormal operation.

Collectively, these activities are called *condition monitoring*. In addition, preventive maintenance services preserve equipment life with:

Lubrication to reduce friction that causes heat, wear, misalignment, or seizure.
Routine cleaning and adjusting done in conjunction with inspection or lubrication, or performed by operators.
Replacement of minor components to reduce chances of more important components failing.

Preventive maintenance should not include major repairs. It is a "detection oriented" activity aimed at uncovering problems before equipment failure and providing sufficient lead time to plan selected work. To avoid confusion, it is best to refer to PM as "preventive maintenance services."

Philosophy

At the outset of the PM program, inspections will identify many needed emergency repairs because a special effort is being made to find problems before equipment fails. If the emergency repairs are made promptly, their volume will soon diminish. Then, as inspections continue, the nature of the deficiencies will gradually change. There will now be more unscheduled repairs and fewer emergencies. Some of the unscheduled repairs will be larger jobs which could be both planned and scheduled. But, make sure a planning capability exists. As the planning capability increases, there will be a further reduction in unscheduled repairs. Many such jobs will meet the criteria for being planned. Thus, the pattern of work generated by PM inspections will change over time. See Figure 3–1. These improvements will be dependent on consistent inspections and testing together with the establishment of an effective planning staff.

Beyond 18 months, the deficiencies yielded by inspection and testing will stabilize, in terms of manpower used, at about 10% emergency repairs and 20% unscheduled repairs. Work that can be both planned

WORK GENERATED BY PM INSPECTIONS VERSUS % OF JOBS OR MAN HOURS USED

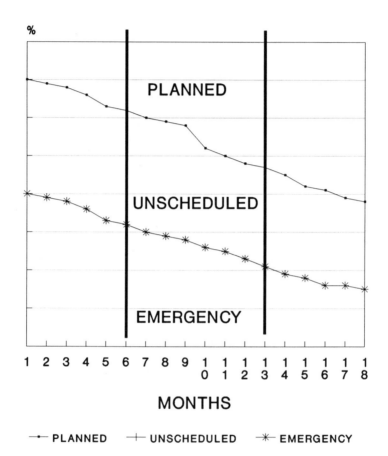

MONTHS

—•— PLANNED   —+— UNSCHEDULED   —*— EMERGENCY

FIGURE 3–1. Within 6 months, inspections and testing will yield mostly emergency repairs, some unscheduled repairs, and little planned work. After approximately 12 months, there will be a significant reduction in emergency work with some increase in unscheduled work and a marginal increase in planned work. By 18 months, emergency work should drop to about 25% of what it was at the start while unscheduled work will drop marginally and planned work will increase substantially.

and scheduled will gradually increase to utilize 40 to 50% of total maintenance manpower. Subsequently, inspections and testing will yield more information on component conditions than on deficiencies. As this happens (approximately 24 months), a periodic maintenance (refer to Appendix E) forecast should be introduced to project future major components replacements (like drive motors). The accuracy of the forecast depends, of course, on the quality of repair history data (component life span) from which forecasts are developed. As component replacements are identified, the most recent PM inspection results should be reviewed to determine the actual condition of the component being replaced. See Figure 3–2.

Scope

The preventive maintenance program should be applied selectively to production equipment, buildings, and facilities. Production equipment should be included in the program based on a high probability of failure and serious consequences of failure. Typically, a one of a kind unit of production equipment whose failure will stop operations is a primary candidate for inclusion in the PM program.

Benefit

Preventive maintenance reduces downtime on production equipment and can maximize its use for operations. If applied to buildings and facilities it ensures that their life is extended.

Objectives

Preventive maintenance services have the following objectives:

1. *Reduction of Emergency Repairs:* Equipment inspection and testing ensure that maintenance is aware of equipment condition and can act to prevent premature failures. By uncovering problems early, emergencies are averted and more deliberate repairs can be conducted.
2. *Reduction of Unscheduled Repairs:* The incidence of unscheduled repairs is reduced. Timely PM inspections identify problems sooner

EQUIPMENT DEFICIENCIES VERSUS JOBS OR % MANHOURS USED

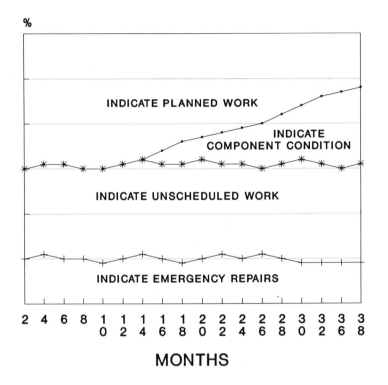

FIGURE 3–2. After approximately 24 months, PM inspections and testing will start to produce information primarily on component condition rather than on equipment deficiencies. Thus, there will be a substantial increase in the amount of planned work developed from inspection and testing.

while they are correctable as adjustments or minor component replacements.

3. *More Planned and Scheduled Work:* Successful PM inspections yield more lead time for planning. As a result, more work can be planned.

4. *Better Manpower Utilization:* As a result of more planned work, productivity on the job is increased because the work was better organized beforehand.

5. *Reduction in Repair Costs:* Emergencies and unscheduled repairs are replaced with more planned work. Because work was organized in advance, labor use is more productive, and less labor is required. Thus, labor costs are reduced.

6. *Reduced Downtime Cost:* Downtime costs 3 times the cost of performing the maintenance work that could have avoided it. By effectively planning and scheduling the job, a 6% reduction in elapsed downtime can be realized. The ability to plan derives from PM.

7. *Preservation of Assets:* Well cared for equipment, buildings, and facilities last longer.

MAINTENANCE ACTIONS

Coordinating Program Elements

Maintenance must ensure that the PM program functions in conjunction with other elements of its program, including:

1. *Work Order System:* A standing work order is normally used to denote PM services because they are routine, repetitive activities.

2. *Planning:* Once PM services are determined, their frequencies established and checklists prepared, the planning related to PM execution is done. On the other hand, deficiencies identified when equipment is inspected often yield work that requires planning.

3. *Scheduling:* PM services for fixed plant equipment are scheduled routinely and repeated at regular, fixed time intervals such as weekly and at four weeks. Mobile equipment PM services are scheduled against variable intervals measured in elapsed operating hours, miles, and so forth.

4. *Information System:* The information system provides feedback on whether or not PM services are carried out on time. It also reports

compliance with the PM schedule as well as changes in labor use, costs, or downtime as a result of performing PM services.

Getting Personnel Organized

As the PM program is set up and carried out, the maintenance superintendent, for example, would conceptualize the overall PM program and delegate its development to his personnel. A general supervisor would, for example, oversee the development, coordination, and execution of the PM program applicable to his area of operation. A maintenance engineer would develop the program scope, scheduling procedures, service frequencies, and so forth. He would also prepare detailed instructions with the help of maintenance supervisors and craft personnel. He would determine the applicability of specific nondestructive testing techniques such as vibration-analysis. Once the program is approved, the maintenance engineer would monitor the program's effectiveness and, as necessary, recommend modifications. Technicians would assist in the development of specific techniques such as oil sampling. Thereafter, they would act as advisors as the program is carried out. Frequently, they would also instruct in inspection, testing, or lubrication techniques. The maintenance planner would schedule services and assist supervisors by monitoring feedback from inspections and testing to identify work that can be planned. Supervisors would be responsible for executing the program in their areas of responsibility. Crew members would carry out services as assigned to them.

PREVENTIVE MAINTENANCE PROCEDURES

General Nature of PM Services

PM services for fixed production equipment are scheduled at fixed intervals such as weekly, biweekly, and monthly. Services are carried out on a routine, repetitive basis. In other words, the service is guided by a checklist each time that it is carried out, making it a routine activity. It is also performed at a regular service interval, making it repetitive. Mobile equipment PM services are based on variable frequencies such as operating hours or miles. All mobile equipment should be included in the PM program. However, small vehicles such as pickup trucks are subjected to less comprehensive services than heavier equipment like loaders. Generally, larger units are scheduled by operating hours whereas small vehicles are scheduled against miles.

Conduct of Services

There should be no repairs made as the PM services are carried out, only cleaning, adjustment, and minor component replacements. Unless such a policy is followed, many inspections and services are never completed. Mobile equipment electrical and mechanical PM services should be conducted simultaneously to reduce servicing time as well as downtime. Services should be guided by a checklist. As the service is carried out, the checklist guides personnel through the service to ensure that nothing is missed. The checklist is a guide. Each checklist should be backed up with a detailed training manual explaining how the service is performed. As necessary, the manual should contain pictures of components so that personnel can see the details.

Required Services

PM services include inspection, nondestructive testing, and lubrication. The principal emphasis should be given to equipment inspection and testing so that the condition of equipment can be monitored and deficiencies can be found. The sooner deficiencies are found, the greater the impact on reducing emergencies. Also, by finding problems sooner, the lead time before the repair is required is longer making it possible to plan more work.

Service Frequencies

PM service frequencies are based on the manufacturers' recommendations, experience, and reliability needs. Service frequencies should be adjusted based on equipment performance. Typically, if a one week fixed equipment inspection interval produces no deficiencies, the interval should be extended to 2 or 3 weeks.

Each type of mobile equipment will have a range of services built on increments of accumulated hours or miles. For example:

- *A service (250 hours):* Routine cleaning, lubrication, and checking of specific critical items.

- *B service (500 hours):* Repeat A service, replace selected filters, take oil sample for analysis, and load-test hydraulic system.
- *C service (750 hours):* Repeat A service, replace selected drive belts, coolants, hydraulic fluids, road test brakes, and test on hill climb.
- *D service (1000 hours):* Repeat A service, make engine compression test, test exhaust emissions, and replace selected seals.

Service Time

The service time is the time required to carry out prescribed services. For fixed equipment PM routes, it includes travel time, performing the service, discussion with operators and supervisors, recording deficiencies, cleaning, adjusting and replacing minor components, and then reporting labor. The total time required for any service should not exceed 2 to 3 hours because the work demands intense concentration. Shorter servicing periods retain interest and avoid prolonged exposure in potentially dangerous working areas.

Downtime

The total mobile equipment service time is, for example, downtime debited against maintenance. It is measured from the time the unit arrives at the garage until the unit is released. It includes cleaning, inspecting, servicing, testing, and minor component replacement. To minimize service time, a mobile equipment maintenance facility should provide bays used exclusively for PM. Extensive repairs should be made in other bays. Materials such as belts, filters, gaskets, and so forth, should be prepackaged before the service begins to save time. Helpers should be used on lesser tasks such as filter replacements in order to keep the craft personnel on the more demanding aspects of the service.

Identification of Services for Fixed Equipment

The standing work order (SWO) is appropriate for control of routine, repetitive preventive maintenance services. A SWO represents a specific preventive maintenance service for fixed equipment. It links the service to the cost-center in which it is carried out:

9069001	Lubrication route 01, cost-center 06
9	Standing work order
06	Cost-center 06
90	Lubrication
01	Route 01

Identification of Services for Mobile Equipment

Standing work orders are not used for mobile equipment. Since each service is unique to a specific unit, servicing codes should refer directly to units. For example:

5101191	A service, haulage truck 011
51	Haulage trucks
011	Unit 011
91	A service (250 hours)

Service Routes (Fixed Equipment)

PM services for fixed equipment should be linked together in routes by cost-center based on types of services and similar frequencies. Every unit in a cost-center requiring a weekly inspection should be placed on the same route, for example. The path between units should be arranged for minimum travel time. Time for discussion with operators and supervisors should be built into routes. Maintenance supervisors should walk each route with the craft personnel who will carry out the service. They should orient them on the service and verify that they understand the PM concept. The checklist should be explained, servicing techniques demonstrated, and the route discussed.

Focus on Cost-Centers

Each PM route should be carried out within a cost-center so that manpower used and costs can be attributed to the cost-center. It also permits the effectiveness of the services to be more readily evaluated.

Central Control

Some PM services are centrally-controlled like lubrication or instrumentation. Central control is more appropriate since these personnel operate plant wide and spend little time in any one cost-center.

Pre-PM Actions by Operations

Once notified of the date of the PM service, equipment operators on all shifts should record any problems that they observe to ensure that they are called to the attention of maintenance personnel.

Actions Required Once Deficiencies Are Known

Ensure maintenance acts immediately to correct deficiencies found by inspection and testing. An excellent motivator is to require them to advise operations first. Operations will then follow up to ensure that all serious deficiencies are corrected. Emergency repairs should be corrected immediately. Unscheduled repairs should be recorded so that they can be performed at the first opportunity. Work requiring planning should be assigned to the planner without delay.

PM Workload

The man-hours, by craft, required for PM should be established. For example, a PM inspection route for fixed equipment performed by an electrician once a month at 3 MH (man-hours) per occurrence requires 36 MH/year. If all such routes were computed, the total MH/year by craft could be determined for all PM services carried out in a given cost-center. As actual man-hours are reported, estimates are verified.

Compliance with the PM Program

Management should receive a weekly PM compliance report. The report should state the percent of services completed versus the number scheduled. Services not completed should be identified so that actions may be taken to reschedule them.

Measuring PM Success

A criteria should be established for monitoring the success of the PM program, including:

1. Reduction in emergency repairs
2. Increased scheduled maintenance

3. Reduction in unscheduled repairs
4. Increased equipment life
5. Extended time between repairs
6. Long-term cost reduction

Keeping the PM Program Up-to-Date

New equipment should be added to the PM program promptly and the adjusted workload checked to ensure that sufficient man-power is available. Similarly, obsolete equipment should be withdrawn and adjustments made in the workload. Equipment that will be phased out should have its services gradually diminished.

Responsibilities

A single person should be responsible for the overall development, administration, and monitoring of the PM program. The maintenance engineer is a good choice. The program should be launched as quickly as possible even if checksheets are not complete or routes have to be adjusted later.

Credibility Is at Stake

Maintenance must produce tangible results like reducing emergencies and not talk about what they are going to do.

LUBRICATION

Objective

Lubrication has the objective of extending equipment life. It is organized and administered the same way as other elements of the PM program. There are lubrication routes, checklists, service frequencies, service times, and so forth. The difference is that performance of the service, not detection of problems, is the main objective. Also, there is a higher degree of potential production operator involvement in checking oil and hydraulic fluid levels at short intervals of every shift or every day. However, operators must be trained if they are to do this effectively.

Use of Lubrication Personnel

Lubrication is often centrally-controlled because the total number of man-hours required weekly in any one area is often less than 20 to 25 hours. Since one man works 40 hours per week, it would not be very productive to assign a lubeman to a single area. Regular mechanics can double as lubricators if central control is not possible. However, the method of organization and control must be based on individual plant circumstances.

Automatic Lubrication Systems

An automatic lubrication system dispenses the correct lubricant at the proper place in the right amount at the proper time. Cost reduction is possible because the system is a substitute for lubrication personnel. However, the system itself must be maintained.

Oil Sampling

Oil analysis reveals the internal condition of equipment as evidenced by microscopic fragments of steel, bronze, iron, and so forth, in the oil. The degree of these fragments reveals, in turn, the amount of wear and signals the state of equipment deterioration. Oil sampling is carried out in a laboratory by lubrication technicians. It is a form of nondestructive testing (predictive maintenance).

NONDESTRUCTIVE TESTING

Nondestructive testing (NDT), also called predictive maintenance, determines equipment condition through the use of testing techniques. These services are administered the same way as other PM services with routes, frequencies, checklists, and so forth. They also produce equipment deficiencies which must be converted into corrective actions. Testing devices must be calibrated to ensure accuracy. Testing compares a current condition with a normal operating condition to identify problems. Some common definitions include:

Infrared Inspection (thermography): The use of infrared imaging equipment to detect incipient failures due to overheating (short-circuits) or lack of heat (plugged boiler tubes).

Magnetic Particle Testing: Detection of cracks, seams, porosity, lack of fusion, or penetration at or near the surface of ferromagnetic materials. A magnetic field is created and it arranges magnetic particles on the surface to reveal the discontinuity.

Vibration-Analysis: Measures the degree of vibration amplitude, displacement, and velocity to compare current internal destructive energy with a normal operating vibration pattern. Results pinpoint problem location and severity.

Ultrasonic Testing: Detection of subsurface flaws (cracks in shafts) and wall thickness measurements (worn piping).

Performing Testing Services

Maintenance should determine whether to perform NDT with its own work force or obtain commercial services. Generally, the frequency of the service and the cost of the equipment are the criteria involved in this decision. If a service is performed regularly it is best to acquire the equipment and develop internal expertise even though the equipment is expensive. If the service is performed infrequently, use a contractor.

Summary

Preventive maintenance is an element of the maintenance program that a maintenance department can ignore—for a time. If they do they will then plead ignorance or overwork and proceed to get buried in emergency repairs. If this happens, it is your plant that will suffer the consequences in unnecessary downtime and a potential threat to profitability. Therefore, use policy guidelines to prescribe how maintenance is to carry out PM and operations to support it. Then, if the PM program is successful there should be fewer emergency repairs, more planned work, better productivity, and lower maintenance cost. Since PM can extend equipment life, capital expenditures for replacements are reduced as well. You must establish indices to signal success.

4

Planning, Scheduling, and Work Execution

What are the essential planning steps?
What type of jobs should be planned?
How can you ensure that planned work is done with the least interruption to operations while making the best use of maintenance resources?

PLANNING, SCHEDULING, AND WORK EXECUTION

Mutually Supporting Actions

Planning major jobs assures productive use of resources, providing the work is scheduled at the best time and, its execution is properly controlled. Definitions to remember are:

Planning: Organizing resources in advance of a major job, so that, upon execution, the work may be carried out more effectively.

Scheduling: Determining and confirming the best time to perform a major job with least interruption of operations and effective use of maintenance resources.

The ultimate purpose of planning is to organize maintenance resources in advance so that when the work is done it can be carried out more effectively. The ultimate purpose of scheduling is to ensure that planned work is done with the least interruptions to operations while making the best use of maintenance resources.

Planned work is done more deliberately. Planning improves the productivity of those doing the work and the quality of their work. It also allows the work to be completed in fewer man-hours and less elapsed downtime. Proper scheduling assures that the best, most convenient time is selected. Thus, there are fewer interruptions. Together, good planning and scheduling achieve better resource use and quality work, reduced downtime, and lasting repairs. While all jobs require some level of planning, the best use of planners is on major jobs that use the most resources. Supervisors would apply informal planning and coordination steps to smaller, more routine jobs. But, to avoid confusion, there should be a criteria for determining which jobs should be planned.

POLICIES

Policies Versus Procedures

The biggest portion of maintenance cost derives from labor and materials used on major jobs. When major jobs are properly planned and scheduled, the benefit is reflected in more productive use of personnel yielding lower labor cost per job and reduced downtime. Ordinarily, the cost of materials installed would be the same for either planned or unplanned jobs. However, such benefits only accrue when operations, maintenance, and supporting departments like warehousing cooperate fully. As a manager, you expect such cooperation. However, the reality is that cooperation is better assured with your guidelines. Consider the following guidelines:

1. A criteria will be developed identifying the type of work that will be planned and scheduled.
2. A priority system will be applied to all planned and scheduled work specifying the importance of each job so that resources can be allocated. In addition, it will establish the time within which the work is to be done so that scheduling goals are established.
3. All planned work will be jointly scheduled with operations to ensure that the best use is made of maintenance resources while ensuring that the work is performed when it least interferes with operations.
4. Upon completion of a scheduling period (usually a week), sched-

ule compliance will be measured to ensure that both operations and maintenance have made a maximum effort to see that the scheduled work has been accomplished. Management will be advised of performance.

5. Planners will focus their efforts primarily on planned and scheduled work. They will not be used to support unscheduled or emergency work except in unusual circumstances and then, only when authorized by the maintenance superintendent.

Guidelines such as these allow maintenance to develop procedures. In turn, common procedures assure understanding and cooperation.

PLANNING CRITERIA

Which Jobs Are Planned

There should be a criteria for determining which jobs should be planned and scheduled. With a criteria, the supervisor knows when the planner should help and the planner agrees. In a typical criteria, the work must be planned and scheduled if:

1. Cost and performance must be measured.
2. A standard must be complied with.
3. Warranty work is being done.
4. Work must be started and completed within a specific period.

Alternately, if any of ten of the following 12 conditions exist, the work will be planned and scheduled:

1. Work not required for at least one week.
2. Duration exceeds one elapsed shift.
3. Requires two or more crafts.
4. Requires crafts not part of the regular crew.
5. Requires two different material sources: stock, purchased or fabricated.
6. Requires coordinated equipment shutdown.
7. Requires supporting mobile equipment, special tools, and so forth.
8. Requires rigging, transportation, and so forth.

9. Job plan necessary.
10. Requires drawings, prints, or schematics.
11. Requires contractor support.
12. Estimated at more than $5,000.

An overly simplistic criteria is not realistic. To require planning if the job will require more than 8 hours or will cost more than $500 is an inadequate criteria.

PLANNERS

What the Planners Should Do

Planners confer with the supervisors on who will perform the work as the plan is being developed. Once the planned work is scheduled, planners should monitor ongoing work and assist supervisors in coordinating work related activities such as on-site delivery of materials. Planners follow a prescribed sequence of tasks during the planning, scheduling, and work execution phases.

1. Prior to initiating planning
   a. Receive requests for planned work and evaluate them against prescribed criteria.
   b. Monitor the long-range forecast to identify work in the immediate future time frame requiring planning.
   c. Confer with supervisors to determine the condition of equipment on which work is due.
2. During planning
   a. Prepare preliminary work orders.
   b. Conduct field investigations of unique jobs. Alternately, apply standards.
   c. Develop a job scope for unique jobs.
   d. Estimate labor by craft.
   e. Prepare bills of materials.
   f. Coordinate with shop planners if shop support is needed.
   g. Assemble drawings, schematics, instructions, and so forth.
   h. Identify mobile equipment, rigging, and transportation needs.
   i. Estimate total job cost.
   j. Establish a target date for scheduling the job.

     k. Obtain job approval.

     l. Determine job priority.

    m. Open the work order in the information system.

     n. Order materials and shop work.

3. Prior to scheduling
    a. Determine the availability of materials and the completion of shop work.
    b. When all materials are available, set up a preliminary schedule with operations.
    c. Tentatively arrange mobile equipment support, rigging, transportation, and on-site material delivery.
    d. Allocate labor by craft to jobs on the preliminary schedule by priority.
    e. Brief supervisors on the proposed schedule.

4. Scheduling
    a. Attend the maintenance and operations scheduling meeting to assist in presenting the schedule. (The schedule should be presented by the maintenance supervisor responsible for carrying it out.)
    b. After schedule approval has been obtained, distribute the approved schedule. (The schedule should be approved by the production supervisor responsible for making the equipment available on which the work will be performed.)
    c. Provide maintenance supervisors with work orders, drawings, and so forth, necessary for them to perform the work.
    d. Confirm mobile equipment support, rigging, transportation, and on-site material delivery that was previously arranged.

5. During job execution
    a. Attend daily coordination meetings and coordinate changes necessary due to delays or job deferral.
    b. Using job information, monitor work initiation and progress.
    c. Assist the supervisor in the coordination of rigging, transportation, use of mobile equipment, or on-site delivery of materials.
    d. Upon job completion, discuss any variances with the supervisor and close the work order.

6. After job completion
    a. Note job cost and performance.
    b. Compare variances with standards or cost estimates where appropriate.

    c. Measure schedule compliance and advise maintenance management.

    d. Observe backlog changes and changes in man-power utilization.

SUPPORTING ROLES

Supervisors' Roles

Maintenance and operations supervisors can make planning successful through proper use of planning.

Maintenance General Supervisors

These supervisors focus the planner's activities on planned and scheduled work and encourage close cooperation between the planner and supervisor. Specific duties include the following:

1. Provide operational control of the planner.
2. Prescribe work to be planned within the criteria for selecting planned work.
3. Specify the time frame within which planned work should be targeted for scheduling.
4. Monitor the long-range forecast and guide the planner in anticipating the planning of forecasted jobs.
5. Approve the preliminary weekly schedule and verify its content.
6. Present the weekly schedule for approval by operations.
7. Monitor the conduct of the schedule and its compliance.
8. Participate in daily coordination meetings and adjust the utilization of maintenance resources should equipment availability change or work be deferred.
9. Monitor cost and performance on major jobs, investigate, and correct significant variances against estimates or standards.
10. Discuss exceptions with the planner and recommend corrective actions.

Maintenance Supervisors

They execute planned and scheduled work based on the approved schedule. During the planning stage, they would confer with the planner to become familiar with job aspects and make recommendations on the method of job execution. Following is a list of typical responsibilities:

1. Assist the planner in field investigations or the interpretation of standards.
2. Confer with the planner on task sequences, use of labor, availability of materials, mobile equipment needs, and so forth, to ensure that the plan is practical.
3. Execute the weekly schedule through crew members and coordination with the planner.
4. Explain significant variances from the plan and recommend changes for future repetitions of similar jobs.
5. Ensure correct, accurate, timely reporting of field data on each job.

Operations Supervisors

They must understand the planning and scheduling sequence and appreciate the benefits of successfully planned work. Their duties include the following:

1. Approve work orders based on estimated cost and timing.
2. Help establish job priorities.
3. Approve the tentative weekly plan presented by maintenance.
4. Participate in the weekly scheduling meeting and approve (or modify) the schedule recommended by maintenance.
5. Make equipment available according to the approved schedule.
6. Participate in daily coordination meetings and, as necessary, adjust the availability of equipment to better meet the schedule.
7. Observe schedule compliance as well as job cost and performance.

PLANNING, SCHEDULING, AND JOB EXECUTION SEQUENCE

Overview

Planning, scheduling steps, and work execution steps are illustrated in Figure 4–1.

PLANNING/SCHEDULING/JOB EXECUTION SEQUENCE

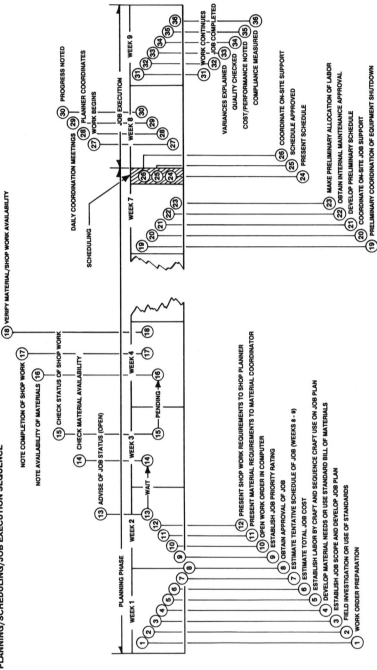

FIGURE 4-1. In the initial stages of planning, planners observe the open work order file to determine which jobs should be started during the weekly period they are planning. They become familiar with the scope and purpose of these jobs. If a job is unique or unusual, they should conduct a field investigation. If the job is performed periodically, they may use a job standard which prescribes details based on previous repetitions of the job.

Steps 1-10: After familiarization with a job, the work order is prepared and used to assemble the job details. These establish man-hour, cost, and time targets against which the job will be carried out. Cost estimates are then made, priorities set, and job approval obtained. The work order is then opened, making it possible to charge material and, subsequently, labor costs against it.

Steps 11-18: The identification and procurement of materials, including shop fabrication, should be accomplished as early as possible to allow adequate time to obtain them.

Step 19: Once the planner has verified that all materials are available and shop work is completed, he/she suggests a target time for equipment shutdown to operations based on equipment condition determined in the latest PM inspection. Operations responds with a time when the equipment could be made available, considering its production needs.

Step 20: The organization of maintenance resources such as transportation, rigging, use of mobile equipment, and so forth, is accomplished after a commitment on equipment shutdown has been obtained from operations.

Step 21: Based on material availability and commitment of production equipment to meet the schedule, the planner coordinates the availability of transportation, rigging, and the use of support equipment from maintenance service departments. Then he/she prepares a preliminary schedule to help coordinate the details of labor allocation or the timing of on-site material delivery.

Step 22: Internal approval of the preliminary schedule is then obtained from the supervisors who will do the work.

Step 23: A preliminary allocation of labor is made to determine whether maintenance can do the work during the period scheduled. A labor utilization report is used to determine the amount of man-power by craft available for scheduled work.

Steps 24-26: The weekly scheduling meeting obtains a commitment from operations to make equipment available and from maintenance to make resources available to perform the work.

Steps 27-34: As the supervisor carries out the schedule, the planner helps by coordinating on-site material delivery, the timing of support equipment, and so forth. Daily coordination meetings help compensate for delays.

Step 35: Information—As work progresses a work order status report provides a picture of accumulated job cost with a summary at completion. The backlog report describes how many estimated man-hours are required to complete the remaining jobs. The closed work order file describes the number of jobs closed during the scheduling period.

Step 36: The percent of jobs completed versus those scheduled demonstrates the overall success of the joint operations—maintenance effort to organize, schedule, and complete the work.

45

Summary

Successful planning, scheduling, and work execution rely on the joint efforts of maintenance, operations, and staff departments like warehousing. You can help assure success with policy guidance, monitoring procedures, and verifying schedule compliance.

5

Labor Control

How many personnel and of what craft does maintenance need?
How should maintenance control overtime and absenteeism?
How can maintenance ensure the effective utilization of its work force?
How is productivity measured?

CONTROLLING MAINTENANCE COST

Labor Control Versus Cost

In many industrial plants, maintenance cost often exceeds 30% of operating costs. Since maintenance controls its cost primarily through the effectiveness with which they install materials, maintenance cost control depends on how well they utilize labor. While other factors like improper equipment operation also affect maintenance cost, they have less influence. Thus, your emphasis on the control of maintenance labor can influence your ability to attain profitability.

FACTORS INFLUENCING LABOR CONTROL

Five Factors

To successfully control the use of labor, it is necessary to:

1. Determine work force size and composition.
2. Measure the utilization of labor.

3. Assess overtime use.
4. Control absenteeism.
5. Improve worker productivity.

Work Force Size and Composition

Each type of work performed by maintenance has a different man power requirement. Therefore, the work load is best measured by first defining the different types of work to be performed. See Appendix D, Maintenance Work Load Definition.

The work load is the essential work to be performed by maintenance and the conversion of this data into a work force of the proper size and craft composition to ensure that the program is carried out effectively.

In turn, work load definitions help to prescribe the type of work and the amount of man-power needed to carry it out. Typically, the work load includes:

- Preventive maintenance
- Emergency repairs
- Unscheduled repairs
- Planned and scheduled maintenance
- Routine activities (like training)

Preliminary estimates are then made of craft man-hours required to carry out each element of the work load. Some portions of the work load can be computed. For example, a PM inspection done every 2 weeks requiring 3 mechanic man-hours per repetition would require 78 mechanic man-hours annually. Similarly, estimated craft man-hours for all PM services would yield a total craft man-power estimate for the entire PM work load.

Similarly, past planned and scheduled maintenance work provides historical man-power use by craft. These can be assembled to provide estimates of current needs. For example, an overhaul due in September requires 210 mechanical man-hours and a drive motor replacement due this month requires 22 electrical man-hours.

Routine activities yield other data. For example, electricians will undergo 65 hours of training annually and one full-time utility man is required for tool repair. When this data is assembled, it provides a good starting point for estimating the number of personnel that will be required. See Figure 5–1.

COST-CENTER 06			CRAFTS			
TYPE OF WORK	MW	ME	EL	WL	LB	TOTAL
PM		1260	1080			2340
SCHEDULED	3060	2160	900	2880	2700	11700
UNSCHEDULED	1620	1620	360	540	540	4680
ROUTINE	1260	1080				2340
EMERGENCY	720	900	180	180	360	2340
TOTAL MH	5400	7200	3600	3600	3600	23400

FIGURE 5–1. By assembling estimated craft man-hours by type of work, an overview of needs can be assembled. (MW = millwright, ME = mechanic, EL = electrician, WL = welder, LB = laborer)

These man-hour requirements are then converted into an equivalent number of craft personnel by dividing each by 1,800 man-hours per person per year. See Figure 5–2.

The result is a preliminary crew of 13 made up of 3 millwrights, 4 mechanics, 2 electricians, 2 welders, 2 laborers, and a supervisor.

Having established the preliminary crew of 13 made up of 5 different crafts, labor utilization data is used to confirm craft man-hours worked against types of work.

A typical distribution is:

Preventive maintenance	10%
Scheduled maintenance	50%
Unscheduled repairs	20%
Routine activities	10%
Emergency repairs	10%

COST-CENTER 06			CRAFTS			
TYPE OF WORK	MW	ME	EL	WL	LB	TOTAL
PM		.7	.6			1.3
PLANNED	1.7	1.2	.5	1.6	1.5	6.5
UNSCHEDULED	.9	.9	.2	.3	.3	2.6
ROUTINE		.7	.6			1.3
EMERGENCY	.4	.5	.1	.1	.2	1.3
TOTAL MEN	3.0	4.0	2.0	2.0	2.0	13.0

FIGURE 5–2. An estimated total of 13 workers would be required to perform maintenance in cost-center 06 with a craft distribution as shown.

Using labor utilization information, the actual distribution of craft man-hours is compared with the desired levels previously listed. For example:

Type of Work	Target	Actual	±
PM	10	3	− 7
Scheduled	50	27	−23
Unscheduled	20	40	+20
Routine	10	10	
Emergency	10	20	+10

As more data is gathered and performance comes closer to the target distribution, the work load can be confirmed. Note the improvement in actual versus target distribution.

Type of Work	Target	Actual	±
PM	10	7	−3
Scheduled	50	45	−5
Unscheduled	20	20	
Routine	10	10	
Emergency	10	18	+8

Supervisors Required

If a traditional maintenance organization with a supervisor and crew is used, the number of supervisors required must be determined so that control of the crew members' activities can be assured. First, determine the number of shifts requiring supervisory coverage annually.

Areas	Annual Shift Coverage	Shifts per Year
3	52 weeks × 5 shifts per week	780
4	52 weeks × 7 shifts per week	1456
	Total Shift Coverage Required	2236

Then, determine the shift coverage capability per supervisor.

$$5 \text{ shifts per week}$$
$$\times (52 \text{ weeks} - 7 \text{ weeks of vacation, holidays, and compensatory time})$$
$$= 225 \text{ shifts per supervisor per year.}$$

Next, divide the total shift coverage required by the shift coverage capability of a single supervisor.

2,236 shifts/225 shifts per year per supervisor = 9.9 or 10 supervisors. Since there are only 7 areas, there should be 7 full-time supervisors and 3 relief supervisors.

Using this data as a guide, the maintenance organization must then decide whether some shifts actually require a maintenance supervisor. For example, a plant with a small crew on each afternoon and night shift might place the crews under the operational control of the operations shift supervisor. However, if this is done, the operations supervisor must be aware of the work that the crew has been assigned and under what circumstances (for example, an emergency) he/she will interrupt their assigned tasks. Similarly, crew members must understand this working relationship.

Labor Utilization

The effective use of labor by category of work is an excellent barometer of the quality of labor control. When adequate man-hours are spent on PM, the result is a reduction in manpower used on emergency and unscheduled repairs with an increase in manpower used on planned and scheduled maintenance. Since planned work is better organized, it is performed more productively. See Figure 5–3.

MAINTENANCE LABOR UTILIZATION REPORT
WEEK 40 ENDING SEPT 30

DEPARTMENT 207

CRAFT	PREV MTCE	SCHD MTCE	UNSC RPRS	EMER RPRS	NON- MTCE	TOTL WKFC
MILLWRIGHT	55	276	102	88	67	588
MECHANIC	46	244	96	66	42	494
ELECTRICIAN	23	122	54	32	46	277
WELDER		72	58	21	66	217
INSTRUMENT	18	32	30	12	45	137
PIPEFITTER		48	12	9	12	81
LABORER		121	6	5	41	173
TOTAL MH	142	915	358	233	319	1967
% DISTRIBUTION	7	47	18	12	16	100

FIGURE 5–3. Man-hours reported against categories of work yields a picture of the effectiveness of labor utilization. In the instance illustrated, 7% of manpower spent on PM has resulted in 47% of manpower used on scheduled work while limiting unscheduled and emergency repairs to 18 and 12% respectively.

A companion piece to labor utilization is the backlog by which maintenance may do the following:

1. Determine whether it is keeping up with the generation of new work. (The backlog does not increase.)
2. Adjust the size and composition of the work force as the work load changes. (A rising backlog for one craft indicates a shortage of personnel while a falling backlog indicates excess personnel. When the backlog is stable for a certain craft, there are enough personnel to keep up with the work.)

Controlling Overtime

Overtime use should be classified by "reason for use" to determine how effectively it is being used and controlled. Concurrently, information on overtime confirms whether or not established overtime approval policies are complied with and overtime is used effectively. See Figure 5–4.

MAINTENANCE OVERTIME SUMMARY
WEEK 40   ENDING 30 SEPT

OCC CODE	DESCRIPTION	1 CALL-IN	2 EMERGENCY	3 SCHEDULED	4 UNAUTHORIZED	5 CONTRACT	TOTAL
01	RIGGER					5.0	5.0
03	DIESEL RPRMAN	10.0					10.0
04	WELDER/MECHANIC			20.0			20.0
05	MECHANIC			35.5			35.5
10	FITTER					3.0	3.0
13	BIT GRINDER						
17	MACHINIST		6.0				6.0
20	ELECTRICIAN		2.0				2.0
25	WELDER				12.0		12.0
37	BOILER MAKER		1.0				1.0
*	*	*	*	*	*	*	
TOTAL OVERTIME MH		98.0	46.0	80.0	14.5	18.0	265.5
% DISTRIBUTION		38.2%	17.9%	31.1%	5.7%	7.1%	100.0 %

FIGURE 5–4. A weekly report of overtime use by craft and reason helps to determine how effectively overtime is being used and controlled. In the report illustrated, 38.2% of overtime was utilized because personnel had to be called back to the plant after the shift. Overtime of 17.9% used was required for emergency repairs which kept personnel on after the shift ended.

Controlling Absenteeism

Poorly controlled absenteeism can be a major detriment to the effective control of labor. Therefore, the same level of control applied to the use of labor must be applied to absenteeism. Typically, an effective absentee report shows absenteeism by reason and craft for a weekly period. See Figure 5–5.

Improving Worker Productivity

Productivity is a measure of the quality of the control of labor. Productivity measures the percent of time that maintenance personnel are able to be at the work site with tools performing productive work. Although actual measurements are most commonly made using random work sampling techniques, the information system should contain indices that allow maintenance to observe "macro" trends in productivity. By measuring the labor cost to install each dollar of material, you can establish whether maintenance is managing the primary means by which they can control costs—the efficiency with which they install materials. As the cost of labor to install each dollar of material increases, it indicates less efficiency in the use of labor.

MAINTENANCE ABSENTEE SUMMARY
WEEK 40 ENDING SEPT 30

DESCRIPTION	PER BUS	UNN BUS	VAC ION	JRY DTY	LVV ABS	INJ URY	NAT GRD	BRV MNT	LTE RPT	DSP ACT	OTH ER	TOT AL
MILLWRIGHT	12	6	80		8	2						108
MECHANIC	4	2	40	2	8			8				64
WELDER	1	3	40			8			2		1	55
MACHINIST	4								4			8
ELECTRICIAN	5	8	80		4	1			2	2	1	103
INSTRUMENT		1					40		1	1	1	44
PIPEFITTER	1	4	40		3	4			2		1	55
TOTAL MH	27	24	280	2	23	15	40	8	11	3	4	437
% DISTRIBUTION	6	5	64	1	5	3	9	2	3	1	1	100

FIGURE 5–5. The absentee report allows the reader to assess the effectiveness with which absenteeism is controlled. Trends from week to week permit compliance with absenteeism policies to be measured. In the instance illustrated, 11% or 51 hours of absenteeism was due to personal and union business.

Summary

Labor control is vital to effective maintenance. It is the primary way that maintenance can control the cost of the work it performs. It starts with finding out how many personnel are needed and then ensuring that they are used effectively. Effective labor use is the result of getting PM done, planning, and quality supervision.

6

Information

What information does maintenance need to manage its activities?
How will they develop this information?
Who should receive what information and what actions should they take?

MAINTENANCE INFORMATION OBJECTIVES

The System

A maintenance information system is the means by which field data is converted into useful information so that maintenance can control its activities. Information to manage the maintenance function must be correctly identified. Thereafter, field data necessary to produce it must be collected and reported. A work order system must exist to focus field data into a data processing scheme which, in turn, produces the required information. The information, once obtained, must then be presented in an appropriate format to those who need it. Finally, those receiving the information utilize it to make decisions and take corrective actions. The success of a maintenance program depends on key people making correct decisions. In turn, the value of these decisions depends on the quality, timeliness, accuracy, and completeness of the information on which they are based.

Information is the basis for taking actions that ensure that maintenance program goals are met. It has the following three objectives:

1. *Determine what work must be done.* The study of repair history data reveals the pattern of failure as well as the life span of specific components to guide planners in scheduling their subsequent replacement. Equipment inspections yield deficiencies on which other, new work is based.
2. *Justify the actions that maintenance must take. Ask why, when, and how much?* Using the cost report, determine the highest equipment repair costs. Investigate specific problems through repair history and then make repair decisions. Observe unit costs. Based on the magnitude of these costs, establish priorities to determine which units will receive attention first.
3. *Confirm the validity of the actions and measure their effectiveness.* What was the use of manpower and of overtime? Were materials used wisely? As a result of observing trends, confirm how effectively the work was done and its cost. Observe the actual costs against the budget and determine the effectiveness of cost-control measures being used.

ESSENTIAL INFORMATION

Types of Information

Maintenance information includes:

Decision-making information to control day-to-day maintenance and determine current and long-term cost and performance trends.
Administrative information to communicate within maintenance and operate the maintenance information system.

Decision-making information embraces of six areas:

1. Labor Control
2. Backlog
3. Status of Major Jobs
4. Cost
5. Repair History
6. Performance Indices

Labor Control

This information portrays labor use and control, work load determination, absenteeism, and overtime use. Labor control impacts maintenance cost. The cost of maintenance is a function of the number of units of equipment maintained and the speed at which they consume repair materials. Emergency repairs accelerate the consumption of materials whereas PM inspections act to slow the rate of material consumption by reducing emergency repairs. PM also makes more planned work possible. In turn, planned work uses labor more efficiently, using fewer hours to accomplish the same work. More planned and scheduled maintenance is more cost efficient. Emergency work wastes manpower. See Figure 6–1.

Work load measurement determines the correct work force size and craft composition to carry out essential work. Procedures to minimize absenteeism and ensure effective use of overtime help ensure the best application of available labor resources.

Backlog

The backlog is the total number of estimated man-hours of work, by craft, required to perform all identified, but incomplete work. It has two functions:

1. Identifying the degree to which maintenance is keeping up with the generation of new work.
2. Adjusting the work force size and craft composition as work loads change.

A common mistake is to consider all identified but incomplete jobs as the backlog. It applies to planned and scheduled work. Emergency work is excluded from the backlog because it must be done immediately. Unscheduled repairs, usually of 2 hours or less, are not estimated and turn over so rapidly that they are excluded as well. Routine PM services are not in the backlog because their repetitive listing would enter and cancel estimated man-hours with such regularity as to constitute a waste of effort. The "backlog" cannot be a listing of work orders. This precludes determining the proper size and craft composition of the work force because a 2 hour job is no different from a 70 hour job. See Figure 6–2.

PM PROGRAM STARTUP
Workload Changes

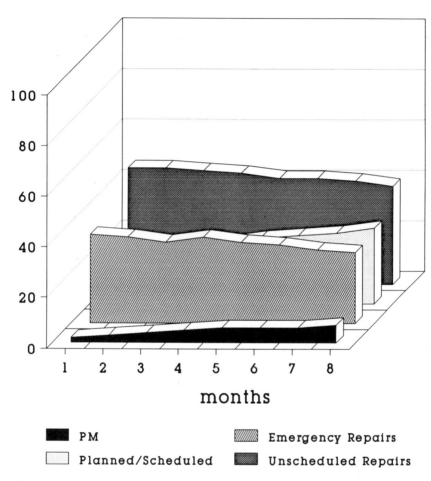

FIGURE 6–1.   An increase in manpower spent on PM acts to reduce the manpower required for emergency and unscheduled repairs while increasing the manpower available for planned and scheduled work.

BACKLOG SUMMARY REPORT
WEEK 40   ENDING 30 SEPT
MAINTENANCE BACKLOG

CRAFT	PRIORITY (90-70)				PRIORITY (69-50)				PRIORITY (49-25)				PRIORITY (24-02)				ALL PRIORITY		
	ST WK	+ WK	- WK	ND WK	ST WK	+ WK	- WK	ND WK	ST WK	+ WK	- WK	ND WK	ST WK	+ WK	- WK	ND WK	ST BL	ND BL	±
MECH	148	13	40	121	271	47	28	290	110	27	12	125	27	12	6	33	556	571	+15
ELEC	96	14	28	82	181	29	66	145	107	31	16	122	42	17	12	47	427	396	-31
WELD	57	10	13	54	201	47	28	220	62	46	21	87	16	21	2	35	336	396	+60
RIGR	128	46	51	123	142	96	57	181	91	28	14	105	81	46	1	126	442	535	+93
FITR	287	49	121	215	108	40	21	127	47	17	12	52	42	12	2	52	484	446	-38
HLPR	147	28	62	113	122	47	21	148	28	30	10	48	17	5	3	19	314	328	+14
TOTAL	1247	128	218	1157	1582	317	417	1482	562	107	172	497	210	98	105	203	3601	3339	-262

FIGURE 6–2. During the week illustrated, the backlog for mechanics increased while the electrical backlog decreased, as did the overall backlog.

Status of Major Jobs

This information provides information on the cost and performance of important jobs such as overhauls, rebuilds, or major component replacements (like engines). These jobs are costly, use considerable resources, and are performed on vital units of equipment. They warrant being managed. Each such job should be planned using a formal work order which establishes job targets for man-hours required, costs, and elapsed completion time. Often, such jobs have standards which prescribe how the work will be performed, what the end product will be, the resources that will be used, and the cost limits. The work order prescribes these goals, and, as work goes on, accounting documents (time card or stock issue card) accumulate labor and material used against the job. The information system provides the cost and performance status from job inception to completion. On completion, cost and performance are summarized and compared with targets. See Figure 6–3.

Cost

Cost control and reduction are the means by which maintenance is often judged. Maintenance costs are built on the expenditure of labor and the consumption of materials. Each unit of equipment, building, or facility should have a unique number so that costs can be accrued against it. Costs also accrue against functions, such as snow removal, custodial work, and grounds maintenance. All costs are further grouped by cost-centers. See Figure 6–4.

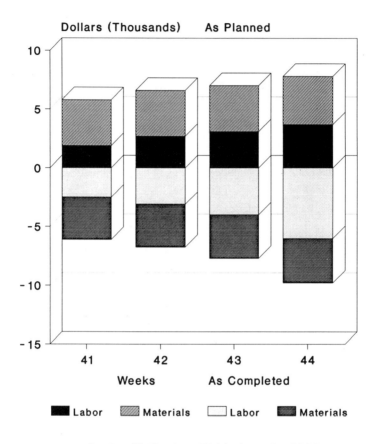

FIGURE 6–3. The illustrated work order shows that the final job cost was greater than the estimate even though material cost was less.

Such cost data is the basis for economic decisions, such as overhauling a unit of equipment, phasing out or replacing a unit of equipment, and standardizing types of equipment based on their cost performance.

COST REPORT
July

COST CENTER - 50061: Section 061
01 WAGNER 150 TON - 021
                    *   *   *
01 WAGNER 150 TON - 022

				CURRENT MONTH			YEAR-TO-DATE			
TP	UNIT	CP	DESCRIPTION	MH	LBR	MTL	TTL	MH	LBR	MTL	TTL	AVG
01	022	10	ENGINE	10	200	150	350	20	400	300	700	100
01	022	20	TRANSMISSION	5	100	50	150	15	300	190	490	70
01	022	30	DRIVE TRAIN	10	200	140	340	30	600	400	1000	143
01	022	40	BED	5	100	70	170	12	240	120	360	52
01	022	50	TIRES	15	300	350	650	35	700	950	1650	235
01	022	60	HYDRAULICS	5	100	75	175	10	200	350	550	78
01	022	70	ELECTRICAL	13	91	210	301	15	300	120	420	60
01	022	80	BRAKES	10	200	320	520	25	500	750	1250	178
01	022	90	FRAME	2	40	20	60	2	40	20	60	9
01	022	99	GENERAL	15	300	10	310	20	400	50	450	64

TOTAL UNIT - 022   90 1631 1395 3026   184 3680 3250 6930   990
===
                    *   *   *
01 WAGNER 150 TON - 023
                    *   *   *
TOTAL COST CENTER - 50061
                    *   *   *
TOTAL MINE

FIGURE 6–4. A typical cost report shows the cost of labor and material to component level of each unit, within a cost-center, monthly with year to date summaries.

Repair History

Repair history is the chronological record of significant repairs made on key units of production equipment. Its analysis reveals patterns of problems which identify corrective actions. Repair history also identifies failure patterns leading to remedial steps. Component life span is also recorded enabling confirmation of periodic actions to replace components. See Figure 6–5.

```
                         REPAIR HISTORY
                CONTINUOUS MINER 024 CM02 400
                      WEEK 8943 TO 9108

FC MWO# SEC UNIT COMP TR MH YRWK  DESCRIPTION OF WORK                SERIAL # IN    SERIAL # OUT OPHRS
-- ---- --- ---- ---- -- -- ----  --------------------------------- ------------   ------------ -----
1       024 CM02 400 EL  8943 RPL TRAM MOTOR OFF SIDE (JOY)          JM-12-112123   JM12-312413  1102
1       024 CM02 400 EL  9001 RPL TRAM MOTOR OFF SIDE (JOY)          JM-12-223335   JM12-112123  1234
1       024 CM02 400 EL  9005 RPL TRAM MOTOR OFF SIDE (JOY)          JM-12-124498   JM12-223335  1286
1       024 CM02 400 EL  9029 RPL TRAM MOTOR OFF SIDE (TRAMCO)       JM-12-344456   JM12-124456  1589
1       024 CM02 400 EL  9108 RPL TRAM MOTOR OFF SIDE (TRAMCO)       JM-12-129032   JM12-344456  1703
1       024 CM02 400 EL  9003 RPL TRAM MOTOR OPP SIDE (JOY)          JM-12-590234   JM12-129032  3307
1       024 CM02 400 EL  9108 RPL TRAM MOTOR OPP SIDE (NAT)          JM-12-452903   JM12-590234  3490
```

FIGURE 6–5. Repair history reveals a pattern of repairs suggesting a specific corrective action.

Performance Indices

Performance indices are more informal and used primarily to spot trends. They include:

Maintenance cost per unit produced
Percentage of man-hours used on emergency repairs
Weekly schedule compliance
Worker productivity
Percentage of supervision time
Cost of labor to install each dollar of material

INFORMATION PYRAMID

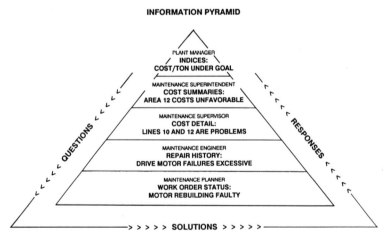

FIGURE 6–6. Questions raised by observing performance indices are answered by the information system. Corrective actions follow.

The effective use of decision-making information and indices reveals management skill. If, for example, the cost per unit of product is excessive, a manager should be able to find out why within the structure of the information system. Typical is the information pyramid. See Figure 6–6.

ADMINISTRATIVE INFORMATION

Internal Administration

Information necessary to run the maintenance department, help communications, produce reference lists, and provide exception information is classified as administrative information such as the following:

An error report showing incorrect time card entries.
A listing of incorrect data entered on work orders.
A listing of stock parts utilized for a specific job.
A listing of overtime hours worked by personnel.
A reference list of employees.

To avoid duplicate administrative information, a criteria should be prescribed and justification required for reports requested. Typical questions to ask are:

Why is the information necessary?
Is the information available elsewhere?
What actions are taken, by whom, with the information?

FIELD DATA—ACCOUNTING SYSTEM— WORK ORDER SYSTEM

A Trio

Field data such as man-hours used or parts consumed are reported by maintenance personnel on accounting documents such as the time card or stock issue ticket. Concurrently, they match the field data recorded with equipment numbers or work order numbers specified on the work order elements used to assign their work. Both the work order data and the accounting data enter the data processing scheme where they are combined to produce information. See Figure 6–7.

WORK ORDER SYSTEM AND ACCOUNTING INTERFACE

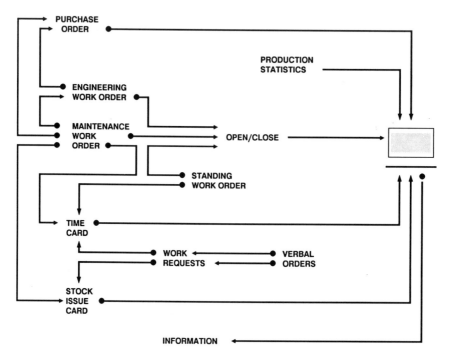

FIGURE 6–7. Work orders identify work to be done (by personnel) as well as equipment, functions, and jobs on which work will be performed (with data processing). As materials are ordered and work is scheduled and performed, field data flows into the system via accounting documents. The result is information.

Field Data

Maintenance struggles to obtain valid field data largely because they have formed bad habits. Supervisors fill out time cards for crew members when they often do not know what crew members actually did. Crew members need a work order to obtain stock materials. Since they are operating under verbal orders, they use yesterday's work order number thus charging materials to the wrong equipment. Look at cost reports. If you see materials installed without labor costs or troublesome equipment with neither labor nor material costs, your maintenance department is among those having trouble obtaining valid field data. The data handling system is seldom the problem. The solution is

training of maintenance workers and vigorous follow up by maintenance supervision.

THE WORK ORDER SYSTEM

What It Does

The work order system is a communications system by which work is
requested, classified, planned, scheduled, controlled, and subsequently directed through a data processing scheme to produce information. If maintenance performs non-maintenance work (like
construction) or uses contractors, the system must be able to control
this work as well. Specific work order elements support different types
of work performed. See Table 6–1.

Typically, about 65% of all jobs performed by maintenance can be
completed by one craftsworker in 2 hours or less. These are unscheduled repairs and few require planning. Therefore, one of the elements
of the work order system should be a simple work request. This allows
personnel to identify equipment and problems easily. By contrast,
major jobs such as an overhaul, rebuild, or major component replacement are costly, important jobs. They must be managed. Therefore,
they should be isolated and their cost and performance measured.
Thus, a detailed, formal work order is usually necessary. The formal
work order establishes the parameters of the job and, in conjunction
with the information system, time card, stock issue card, and so forth,
collects data against that job. The status is recorded as the job progresses and cost and performance are summarized when the job is finished.

Emergency repairs are carried out immediately. Therefore, verbal
orders must be included in the work order system. But, their use

TABLE 6–1   Work Order Element Versus Type of Work

Work Order System Element	Type of Work
Maintenance work order	Planned and scheduled work
Work request	Unscheduled repairs
Verbal orders	Emergency repairs
Standing work order	Routine, repetitive actions
Engineering work orders	Non-maintenance work

should not inhibit control of work or deny important information like repair history. Organizations that disallow the use of verbal orders are being unrealistic. Invariably, their work order systems are inadequate or they cannot muster the internal discipline to successfully utilize verbal orders.

Standing work orders should only be used to describe a function, never a unit of equipment. When a unit of equipment is designated by a standing work order (SWO), up to 40% of its repair cost is lost or unidentified because SWOs are seldom assigned to costly components. Repair history is difficult to collect for the same reason.

A WORD ABOUT PACKAGE PROGRAMS

Bewildering Choices

There are over 100 "package" maintenance programs available. Package programs offer quick installation and, in theory, immediate use after program files are loaded. Unfortunately, it is not that simple. Selecting a package should never be left to maintenance alone. Ensure that accounting, inventory control, purchasing, and data processing are involved. No matter how simple the program, it will soon impact some or all of these departments. Assemble the group and have maintenance sell it to them. If they can, get it. Some packages are fully integrated, meaning that to use the maintenance portion you must also purchase a new inventory control program or even a general ledger. If existing programs other than maintenance are in good shape, write communications software to link what you have with what you need rather than purchase programs you do not need.

SYSTEM IMPLEMENTATION

People Problems

Expect problems with people during information system implementation. The biggest problem will be with older supervisors who do not use the programs effectively, often because they do not want to. It follows that their crew members will emulate this behavior. When this happens, supervisors may have surrendered work control to the planner who, because he/she can use the computer with facility, produces the work orders for the supervisors to hand out. Some excellent systems are not very helpful because few use them. Often, the only personnel who can use them are planners or maintenance engineers. Such behavior does little to help control work, enhance productivity, and ensure the performance that you require of maintenance.

WHO NEEDS WHAT INFORMATION

Ultimate Purpose

Putting the right information in the hands of the right people is the ultimate purpose of the maintenance information system. This includes information about maintenance needed by the plant manager

TABLE 6–2

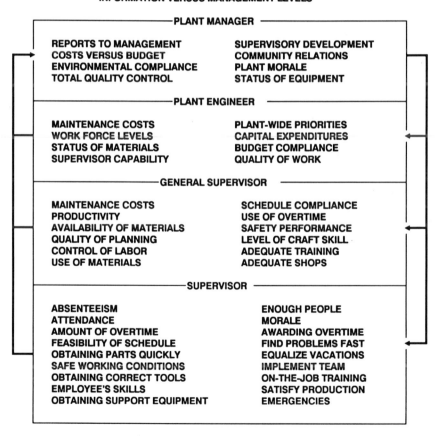

INFORMATION VERSUS MANAGEMENT LEVELS

PLANT MANAGER

REPORTS TO MANAGEMENT	SUPERVISORY DEVELOPMENT
COSTS VERSUS BUDGET	COMMUNITY RELATIONS
ENVIRONMENTAL COMPLIANCE	PLANT MORALE
TOTAL QUALITY CONTROL	STATUS OF EQUIPMENT

PLANT ENGINEER

MAINTENANCE COSTS	PLANT-WIDE PRIORITIES
WORK FORCE LEVELS	CAPITAL EXPENDITURES
STATUS OF MATERIALS	BUDGET COMPLIANCE
SUPERVISOR CAPABILITY	QUALITY OF WORK

GENERAL SUPERVISOR

MAINTENANCE COSTS	SCHEDULE COMPLIANCE
PRODUCTIVITY	USE OF OVERTIME
AVAILABILITY OF MATERIALS	SAFETY PERFORMANCE
QUALITY OF PLANNING	LEVEL OF CRAFT SKILL
CONTROL OF LABOR	ADEQUATE TRAINING
USE OF MATERIALS	ADEQUATE SHOPS

SUPERVISOR

ABSENTEEISM	ENOUGH PEOPLE
ATTENDANCE	MORALE
AMOUNT OF OVERTIME	AWARDING OVERTIME
FEASIBILITY OF SCHEDULE	FIND PROBLEMS FAST
OBTAINING PARTS QUICKLY	EQUALIZE VACATIONS
SAFE WORKING CONDITIONS	IMPLEMENT TEAM
OBTAINING CORRECT TOOLS	ON-THE-JOB TRAINING
EMPLOYEE'S SKILLS	SATISFY PRODUCTION
OBTAINING SUPPORT EQUIPMENT	EMERGENCIES

and operations. In addition, different personnel within maintenance require information that coincides with their responsibilities or work they perform. See Table 6–2.

Summary

Maintenance information includes decision-making information to control maintenance and administrative information for internal communications. Focus on decision-making information. Make certain the essential information is available and provided to the right people. Screen the work order system to ensure that it works in concert with the accounting system. Make certain the field data that drives the system is complete, accurate, and timely.

7

Maintenance Engineering

What is the purpose of maintenance engineering?
How does it improve maintenance performance?

UNDERSTANDING MAINTENANCE ENGINEERING

Objective

Maintenance engineering has the objective of ensuring the reliability and maintainability of production equipment. Equipment reliability is achieved by verifying the quality of the maintenance program and its execution. Maintainability is achieved by ensuring that equipment installation and modification are carried out properly so that maintenance can, thereafter, maintain the equipment effectively.

Key Activities

To meet their objectives of assuring equipment reliability and maintainability, maintenance engineers must do the following:

Develop and monitor programs (like PM).
Establish standards for major jobs (like overhauls).
Monitor the conduct of the maintenance work to determine if standards are complied with.

Observe program execution to determine if work is carried out on schedule.

Assess organizational performance to determine whether organizational change is needed.

Evaluate new installations for maintainability.

Approve equipment modifications.

Assess information quality.

Prescribe new or remedial training.

Based on their observations, they recommend improvement actions to the maintenance manager.

MAINTAINABILITY

Equipment Installation

There is concern that new installations may be performed without due attention to future maintainability. As equipment is being installed, the maintenance engineer should observe the work to ensure that equipment will be maintainable once it is commissioned. For example, a contractor might locate equipment too close to an existing wall making it difficult to inspect or lubricate. When new equipment is installed, the maintenance engineer should be included in the commissioning team. He/she should have the same influence in making maintenance (or a contractor) produce wiring diagrams, for example, as the operating superintendent has in requiring additional work if the equipment fails to yield quality product.

Modification

Equipment modifications must be necessary, feasible, properly engineered, and, since they are not maintenance, correctly funded. Many are capitalized. There is a tendency in some operations to justify equipment modification based on potential improvement in equipment performance. The initiator may even attempt to bypass sensible feasibility checks on his modification. Often, modifications are performed by maintenance supervisors ignorant of the possible consequences.

For instance, one maintenance supervisor carried out modifications on several similar units without consulting anyone. None were modified in the same way nor was there any documentation. Stocked parts to fit the original design matched none of the modified equipment. Thus, future repairs took much longer creating unnecessary downtime. Modifications, if not monitored, can be harmful. If not screened, they may consume significant labor and threaten a shortage of labor required for basic maintenance. Assessment of possible modifications by the maintenance engineer can avoid future problems.

UTILIZING MAINTENANCE ENGINEERING

The Assessment Function

Maintenance engineering impacts performance by assessing the organization, its program, and the environment in which the program is carried out. The assessment leads to corrective actions.

Organization

The maintenance organization must fit the production circumstances of the plant. For example, if quick response is necessary, the lethargy of a centrally-controlled organization should be avoided. Similarly, if production continuity depends on periodic shutdowns, a strong planning and scheduling element is necessary. The craft organization with a crew of millwrights who cannot cope with minor electrical problems serves operations poorly. Downtime accrues while an electrician is sought. A poorly-trained electrician wastes time with "trial and error" solutions. Through continuous field observation, the maintenance engineer can assess organizational responsiveness and make recommendations for corrective actions.

ASSESSING THE MAINTENANCE PROGRAM

Program

The maintenance engineer is interested in the completeness of the overall maintenance program, including the continuity between its elements, the quality of procedures, and the degree to which personnel carry out the program effectively.

Preventive Maintenance

The preventive maintenance program is a special domain of the maintenance engineer. He ensures that the inspection, testing, and lubrication aspects are appropriate and being carried out effectively. He should develop the program and train maintenance personnel in its use.

Planning and Scheduling

Planning and scheduling are also areas of interest to maintenance engineering. After several years of operation, planners will have built a data base reflecting the actual life span of most major components. This data is used to forecast the timing of the next replacement. The maintenance engineer ensures that the forecast and the PM program are properly coordinated so that components are not replaced prematurely. The maintenance engineer also prescribes repair techniques, tools, and materials for these periodic maintenance tasks.

Information

The use of information has a significant impact on the overall organization effectiveness. The maintenance engineer should review the content of the information system as well as the quality and timeliness of the field data from which it is derived. Of special importance are cost, performance, and repair history information. Cost information identifies units with high costs. This leads to examination of performance information like cost per operating hour. In turn, repair history is examined to uncover probable causes. The maintenance engineer then pinpoints the exact problem with a field investigation. This ensures realistic recommendations to correct problems.

Material Support

Maintenance depends on quality material. It is the maintenance engineer who verifies the durability of repair materials, the quality of work in rebuilding components, or the proper quantities of spare parts.

Selection of Equipment

Equipment selection, especially used equipment, should have input from maintenance engineering to avoid poorly performing or difficult to maintain units.

THE PLANT ENVIRONMENT FOR MAINTENANCE

Environment

A maintenance organization and a maintenance program operate best in a supportive environment when operations is responsible for the effective use of maintenance services. In addition, material control, accounting, and data-processing personnel must appreciate maintenance dependence on their services and deliver them effectively. When a cooperative environment exists, the maintenance organization can be efficient and its program will be carried out effectively. The maintenance engineer is in a position to observe how well the plant environment supports maintenance. Everything from the willingness of operators to perform PM related tasks to the dedication of the purchasing agent in tracking down urgently needed spare parts can be observed by the maintenance engineer. The engineer's ability to keep observations in perspective makes them valuable. It would be prudent for any maintenance manager developing a new program to include the maintenance engineer in the development team. Similarly, once the program is operational, it would be equally prudent to count on the maintenance engineer for a continuous evaluation of maintenance program performance.

ORGANIZATIONAL ASPECTS

The Engineer

In a large maintenance organization, the maintenance engineer should be a bona fide engineer. The discipline acquired in an engineering education is a definite plus to the well-ordered nature required of maintenance engineering. When field experience is added, capabilities are reinforced. It does not disqualify the non-degreed candidate who must work harder to achieve some of the skills that the engineer already possesses. This can distract from essential duties. The maintenance engineering function must enjoy organizational prominence. The maintenance engineer should report directly to the maintenance manager. A maintenance work force exceeding 50 personnel should have a formally organized maintenance engineering group. That group might include technicians whose exclusive specialty might be oil analysis or infrared testing. A small work force can perform the maintenance engineering functions using supervisors, planners, and craft personnel. They are usually successful because they see issues in practical terms and relate well with members of the work force. Every maintenance organization performs aspects of maintenance engineering. It remains to determine the degree, identify necessary functions, and organize to carry them out.

Summary

Maintenance engineering develops the vital activities of preventive maintenance, work standards, planning, and information. It assures the maintainability of newly installed and modified equipment. Also, it impacts organizational performance and the environment of supporting activities like material control. It is an activity that the maintenance organization aspiring to excellence cannot be without.

8

Non-maintenance Work

How should non-maintenance work like equipment installation or modification be developed, organized, and carried out?
To what degree can maintenance perform non-maintenance work without it interfering with their basic maintenance program?

INTRODUCTION

Engineering Project Work

Non-maintenance work includes construction, equipment installation,
modification, and relocation. It may be performed by the regular maintenance work force, a crew dedicated to this work exclusively, or by a
contractor. Contractors may supplement the maintenance organization
or perform the work alone. Regardless of how the work is accomplished, maintenance will have to maintain the equipment installed or
modified as well as facilities constructed. Thus, their involvement in
project approval and accomplishment is essential.

Common Problems

Maintenance may have problems performing non-maintenance work if
the work is not well identified, properly assessed, and prepared for.
Typically, the work may be initiated without proper engineering assessment. Thus, the work is more difficult to accomplish and the equipment, once installed or modified, could be harder to maintain. The

work may be expensed rather than capitalized causing the maintenance budget to be overspent. Manpower intended for maintenance work may be used thus reducing its availability for basic maintenance. A maintenance department that is required to perform non-maintenance work along with maintenance work must ensure that its labor resources are properly allocated. While maintenance may be aware of this, plant management and operations frequently are not. For example, at a large paper mill, operations required equipment modifications to improve output and product quality. With this justification, management encouraged these projects. Maintenance was required to start work before engineering could verify that the project was necessary, feasible, and correctly funded. Projects were undertaken before process engineers could assess their impact. Some modifications were extensive enough to require capital funding but, instead, were expensed. Basic maintenance was not being done. Emergency repairs and "projects" became the main activities of maintenance. Soon, the modifications became a last ditch effort to keep equipment running.

Policy

Policies should exist for determining the necessity and feasibility of non-maintenance work and how it should be funded. Policies ensure that both process and maintenance engineers can judge the proposed project before it is performed by maintenance. In addition, accounting guidelines would be applied to ensure that the project is not expensed when it warrants capital funding. The management policy should be specific in the use of maintenance resources. A provision should be stated, for example, in which maintenance is precluded from committing labor to projects until all basic maintenance work has been covered. A typical policy might be:

> Non-maintenance projects, such as construction, equipment installation or relocation, and modifications, will be reviewed to ensure that they are necessary and feasible. They will be subjected to funding criteria before being engineered. If work is to be performed by maintenance, no maintenance resources will be allocated until the basic maintenance work load has been carried out. If maintenance cannot support an approved project, the work may be contracted, subject to current labor contract agreements.

Necessity

Operations is justifiably keen on achieving better equipment performance. Yet, in their eagerness to improve performance, they may overlook the need to ensure that their desired project is sound. Thus, a first step is to ensure that the project is really necessary. Often, projects which appear necessary at the operational level prove unnecessary when examined in the larger context. For example, a general supervisor went to considerable trouble to justify an equipment modification only to learn that the whole production line would be replaced within a year.

Feasibility

An operations supervisor may, for example, have an unrealistic sense of urgency about a pet project and negotiate directly with a maintenance supervisor about getting started. In their eagerness to please, maintenance supervisors may initiate the work without solid preparation. The following results are predictable:

1. The modification is an undocumented "one of a kind" job in which stocked parts no longer fit the changed equipment.
2. Subsequent trouble shooting is difficult because the wiring or controls were changed but not documented.
3. The changed equipment now impacts other related equipment unfavorably causing other problems.
4. In turn, process engineers, ignorant of the change, waste time investigating unrelated problems.

These situations are more common than managers realize. Projects that are worthwhile should not be blocked. Instead, sensible guidelines should be provided to preclude these projects from being carried out based only on informal discussion at the supervisor level.

Allocation of Labor

In many maintenance work forces, labor is controlled by first line supervisors. These supervisors may not realize the need for limiting non-maintenance. Therefore, it is prudent to establish allocation guidelines advising supervisors how to allocate labor to this activity.

Maintenance Qualifications

Assess the backgrounds of personnel to determine if they understand the need to balance their support of maintenance versus non-maintenance. For example, as construction of a new plant was completed, maintenance supervisors and crews were recruited from the contractor. Supervisors and workers job titles corresponded roughly with the millwrights, welders, pipefitters, and electricians needed. However, within 6 months, the maintenance program was in trouble because these personnel preferred, encouraged, or solicited "projects" in lieu of the basic equipment inspection, diagnosis, running repairs, and scheduled maintenance required. Their previous experience had not prepared them for this.

Maintenance Engineer

The maintenance engineer is responsible for the maintainability and reliability of equipment. The engineer must ensure that any new facility construction permits easy access to the equipment inside so that it can be maintained. He/she must determine whether new equipment installations are configured for maintainability. Also, he/she must require maintenance instructions, drawings, spare parts lists, and so forth, for the new equipment. If equipment is to be relocated, the engineer must ensure that the end result yields a placement that permits easy maintenance and that the equipment has not been substantially altered in the moving process. If equipment is to be modified, the maintenance engineer must ensure that the changes proposed do not detract from the desired levels of reliability and maintainability. All modifications must be documented and standardized. Regardless of the non-maintenance work performed, the maintenance engineer should verify that the end result is able to be properly maintained.

Process Engineer

The process engineer ensures the continuity of equipment changes within a specific production activity. Therefore, any proposed modifications should be screened by him/her. The process and the maintenance engineers should collaborate on changes needed based on desired equipment performance.

Workload

Generally, a maintenance work force that has a separate part of its organization dedicated to non-maintenance work can sustain a specific level of this work. Occasional work load peaks can be handled by deferring work, using supplemental contractors or temporarily transferring personnel with similar craft skills from the regular work force. However, when the regular maintenance work force performs both maintenance and non-maintenance work, there must be a balance in work force use. Generally, maintenance can sustain about a 15 to 17% commitment of its work force to non-maintenance work on a continuous basis. A typical distribution is:

Preventive maintenance	9-11%
Unscheduled repairs	12-15%
Emergency repairs	8-10%
Routine maintenance activities	10%
Planned and scheduled maintenance	40%
Non-maintenance work	15-17%

Non-maintenance work should be planned to ensure that it goes smoothly and scheduled so that resources can be properly allocated.

Work Organization

An engineering work order (EWO) should be used for overall control of non-maintenance work. The EWO initially identifies the proposed work. Subsequently, as it is assessed and found feasible, the EWO guides the project through the funding approval steps. Once approved, the EWO is linked with the maintenance work order (MWO) if maintenance will do the work, or with purchase orders if a contractor will perform it. If both work together, the EWO must link with both the MWO and the purchase order. See Figure 8–1.

Scheduling

Non-maintenance projects, like major maintenance work, should be forecasted as far in advance as practicable. This permits maintenance to anticipate the need for its labor resources. Maintenance can then allocate resources by priority to maintenance versus non-maintenance projects.

PROJECT PLAN

WORK ORDER #: 716820   DESCRIPTION: INSTALL 250 FOOT CONVEYOR

MWO DESCRIPTION	WK17	WK18	WK19	WK20	WK21	WK22	WK23	WK24	WK25
WO 512806									
ANCHOR HEADS TAIL PULLEYS	▬								
WO 514706									
INSTALL ELECTRICAL POWER TO AREA		▬							
WO 526307									
INSTALL ROLLERS		▬							
WO 533661									
INSTALL BELTING AND ALIGN ROLLERS			▬▬						
WO 534602									
INSTALL ELECTRICAL CONTROLS			▬				▬		
WO 534814									
INSTALL LUBE SYSTEM			▬				▬		
WO 534892									
CONDUCT SYSTEM TEST									▬

FIGURE 8–1.   The EWO provides the project engineer with overall control of project tasks performed by either maintenance or a contractor, or both.

Summary

Reasonable guidelines are necessary to screen non-maintenance work to ensure that it is necessary, feasible, and properly funded. Thereafter, maintenance engineers should verify that the end result will be maintainable. Process engineers should check equipment changes to determine that process continuity is preserved.

9

Shop Operation and Services

What activities do shops perform and how are they best controlled?
How is component rebuilding best controlled?
What support services are normally available and how should they be controlled?

SHOP OPERATION

Shop Functions

Shop facilities, such as machine or fabrication shops, welding shops, carpentry, paint, and plumbing shops, provide support for field maintenance work as well as for non-maintenance work such as equipment installation.

Shop Work Control

Shops are not normally exposed to the level of unscheduled or emergency repairs that field units are. Shops are typically organized into work stations to facilitate internal control. They are able to plan the majority of their work. The use of work orders for control is higher than with field units. Also, fewer verbal orders are used. The main work control problem is the need to respond to unscheduled requirements of field maintenance units. At times this can be disruptive of shop activities.

Use of Work Orders

Work orders would normally be used for jobs requiring work in several work stations. Alternately, short jobs should have simple work request and control procedures. The number of short jobs can be limited by manpower allocation. Generally, standing work orders are used when parts are being manufactured or components are being rebuilt.

Emergency Work

When a field maintenance unit requests emergency work, require them to determine which of their scheduled work is to be interrupted to accommodate their emergency work. This causes them to carefully consider whether the emergency really exists.

Parts Manufacture

A "make or buy" decision should precede the decision to manufacture spare parts in shops to replenish stock in lieu of purchasing them. The unit cost of shop manufacture should be verified as part of this decision. Economic quantities must also be observed since the machine "set up" cost can be prohibitive. Accurate drawings are necessary before attempting the remanufacture of parts. Verify that manufactured spare parts are placed in inventory so that their issue will be controlled using established procedures. Quality control should be incorporated into repair history to assess the durability of manufactured parts. The rate of parts manufacture should be driven by the inventory control system with attention to economic quantities and competing activities for machine time.

Component Rebuilding

The rebuilding of components is an essential part of cost-control for any operation in a remote location or dependent on a distant (or foreign) manufacturer for spare parts resupply. There are essential control steps that must exist before component rebuilding can be done successfully:

1. Damaged or worn components must be tagged by maintenance identifying the unit of equipment from which they were removed.

2. These worn components must then be moved to a collection point from which the warehouse or shop will retrieve them.

3. The worn or damaged components should then be taken to a classification point where they are cleaned and evaluated by shop personnel. Some may be damaged beyond rebuilding.

4. Based on the maintenance forecast for the future consumption of each component, they are rebuilt in economical quantities and returned to warehouse stock.

5. The cost of rebuilding includes labor used, parts consumed, machine time, and shop overhead. These unit costs should be considered in the "make or buy" decision.

6. Rebuilt components should be returned to stock and issued at the previously determined unit value or as prescribed by local accounting procedures.

7. A method of tracking performance should be included in repair history to track how well each component does.

If components are rebuilt in commercial shops, performance tracking is essential especially if the same component is rebuilt by several vendors.

Planning and Scheduling Meetings

Field maintenance personnel should include shop personnel in planning and scheduling meetings. Thereby, shops will be appraised of field needs and can better meet schedules.

Coordination of Shop and Field Work

"Walk-in" shop work can be very disruptive. There should be procedures for its control. Field planners should ask supervisors to route shop work through them prior to delivering it to the shop.

"Government" Work

Many shop man-hours can be spent in unauthorized work such as making trophies for retiring personnel. Managers often consider it incidental and may even encourage it. However, most change their minds when they find out how much is really done and what it costs.

SUPPORT SERVICES

Service Functions

Support services include transportation, rigging, and lifting services (cranes) as well as clean-up activities. Maintenance organizations may also provide power-generation, heating, steam-generation, or hoisting services.

Services Performed by Shops

Plumbing shops, pipefitting shops, paint shops, and welding shops perform field work in addition to regular shop work. Plumbing shops perform repair services as well as installations while some personnel follow inspection routes moving from boilers to water heaters to personnel toilets. Pipefitting personnel perform similar functions but are more likely involved in installation work rather than repair. Painters focus on specific areas or equipment requiring periodic paint protection.

Control of Services

Transportation and other services requiring mobile equipment, like mobile cranes, are usually controlled by a dispatcher. While many such services are arranged for in advance, the dispatcher controls the exact timing of the use of equipment on each job. With the exception of major jobs, equipment seldom remains on the job site waiting when it can be used productively elsewhere and brought back when needed again.

Field maintenance supervisors and planners should anticipate their need for support equipment and try to request it at least a week in advance. Attendance at scheduling meetings by shops helps to allocate resources against priorities. Daily coordination meetings permit the adjustment of the use of support equipment to preclude unnecessary idle time when delays are encountered. For example, a maintenance planner at a copper concentrator requested transportation, rigging, and mobile crane support for a major conveyor drive motor replacement. Because he had failed to coordinate the exact timing of the shutdown of the equipment, 16 service personnel and their equipment waited idly for over 45 minutes while operations switched to by-pass equipment so that the work could be done.

Summary

Efficient shop support and services are essential ingredients in the success of the overall maintenance program. As a manager, be aware that field and shop activities may, in some maintenance departments, operate as separate entities. By identifying the needs and capabilities of each, you can bring the planning, scheduling, and coordination of their tasks closer together. Thus, you can influence their contribution to your profitability objectives.

10

Material Control

How can maintenance best utilize the services of purchasing and warehousing?
What responsibilities do purchasing and warehousing have in supporting maintenance?
How is the manufacture of spare parts and the rebuilding of components best carried out by maintenance?

MAINTENANCE MATERIAL CONTROL

Definition

Maintenance material control is the effective use of inventory control and purchasing services to ensure that the materials required are made available, on time, and in proper quantities to carry out essential work. In turn, maintenance shops manufacture spare parts or rebuild components in response to needs for a continuous supply of these parts to the warehouse. Thus, mutual cooperation between maintenance, purchasing, and warehousing is essential. Inventory control and purchasing are not normally maintenance functions. Generally, they are accounting functions but are nevertheless vital to maintenance.

Working Relationships

Successful material control for maintenance depends on mutual understanding of the services offered and performed by each group. Maintenance workers must be able to identify stock materials and supervisors or planners must provide adequate lead time when ordering purchased materials. In turn, purchasing and warehousing must appreciate how maintenance operates and support them with a high level of service. Similarly, when maintenance shops manufacture spare parts or rebuild components they must do so in economic quantities against schedules that assure a continuous supply of the required parts. See Figure 10–1.

Roles

Each level of the maintenance organization makes a contribution to effective material control. For example, the superintendent must review stockage levels to identify and remove obsolete parts or stock new parts required. General supervisors should ensure that disabled equipment is not "cannibalized" to fix other equipment. Supervisors must ensure that procedures are followed to obtain stock parts or order purchased materials. Planners should prepare standard bills of materials for forecasted, major periodic maintenance tasks such as overhauls to facilitate procurement. This coordination must work both ways. Just as maintenance must use material control services effectively, a purchasing agent or a warehouse supervisor must have a clear understanding of the overall maintenance program and the material-related roles played by key maintenance personnel.

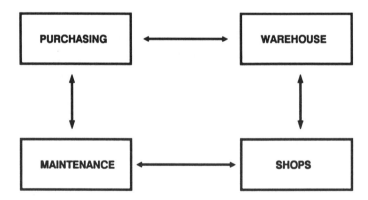

FIGURE 10–1.   A network of mutually supporting activities exists between purchasing, warehousing, shops, and field maintenance units.

MAINTENANCE MATERIAL NEEDS

Maintenance Requirements

Maintenance must be able to obtain the correct materials in the proper quantity at the time that they are needed. Therefore, they must identify materials and quantities needed and the dates that they are needed.

Stock Materials

Average maintenance workers have difficulty identifying stocked parts. Typically, they must select parts from a list or use the computer. Often, crew members may not understand the procedure or find it troublesome. Then the supervisor must get involved. When this happens, the supervisor, whose principal job is supervision, becomes, by default, a clerk. Crew productivity then suffers without his/her control.

Visual parts identification is an effective solution once set up and training is provided. See Figure 10–2. Alternately, equipment specifications programs allow crew members to use the computer to select the equipment type and component. The computer then responds with parts, drawings or diagrams. With system integration, direct access to inventory part listings or engineering drawings is possible. Fully-integrated systems allow the user to add the quantity needed to the part identified, specify the work order number, and order materials. Parts are then withdrawn from stock to be picked up by the requestor.

Major Components

Periodic maintenance includes the replacement of major components such as drive motors, engines, transmissions, and so forth. Repair history includes information on the life span of most major components. Thus, the expected component life is known. This allows its future replacement to be forecasted. Data on previous component replacements also identifies standard stocked parts that were used. Since the approximate timing of the component replacement is known, materials may be ordered in advance, simplifying both maintenance and material control.

TYPICAL MANUFACTURER'S PARTS CATALOG PAGE

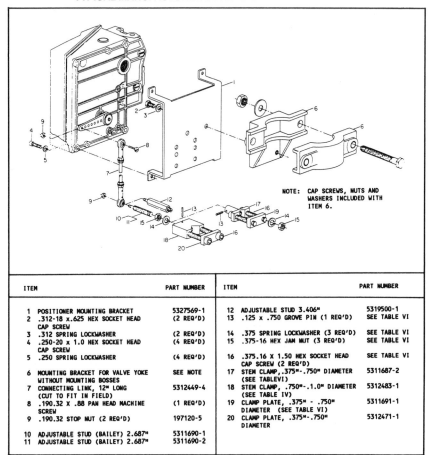

NOTE: CAP SCREWS, NUTS AND
WASHERS INCLUDED WITH
ITEM 6.

ITEM		PART NUMBER	ITEM		PART NUMBER
1	POSITIONER MOUNTING BRACKET	5327569-1	12	ADJUSTABLE STUD 3.406"	5319500-1
2	.312-18 x.625 HEX SOCKET HEAD CAP SCREW	(2 REQ'D)	13	.125 x .750 GROVE PIN (1 REQ'D)	SEE TABLE VI
3	.312 SPRING LOCKWASHER	(2 REQ'D)	14	.375 SPRING LOCKWASHER (3 REQ'D)	SEE TABLE VI
4	.250-20 x 1.0 HEX SOCKET HEAD CAP SCREW	(4 REQ'D)	15	.375-16 HEX JAM NUT (3 REQ'D)	SEE TABLE VI
5	.250 SPRING LOCKWASHER	(4 REQ'D)	16	.375.16 X 1.50 HEX SOCKET HEAD CAP SCREW (2 REQ'D)	SEE TABLE VI
6	MOUNTING BRACKET FOR VALVE YOKE WITHOUT MOUNTING BOSSES	SEE NOTE	17	STEM CLAMP,.375"-.750" DIAMETER (SEE TABLEVI)	5311687-2
7	CONNECTING LINK, 12" LONG (CUT TO FIT IN FIELD)	5312449-4	18	STEM CLAMP, .750"-.1.0" DIAMETER (SEE TABLE IV)	5312483-1
8	.190.32 X .88 PAN HEAD MACHINE SCREW	(1 REQ'D)	19	CLAMP PLATE, .375" - .750" DIAMETER (SEE TABLE VI)	5311691-1
9	.190.32 STOP NUT (2 REQ'D)	197120-5	20	CLAMP PLATE, .375"-.750" DIAMETER	5312471-1
10	ADJUSTABLE STUD (BAILEY) 2.687"	5311690-1			
11	ADJUSTABLE STUD (BAILEY) 2.687"	5311690-2			

FIGURE 10–2. An "exploded" drawing of major components allows personnel to correctly identify parts. Most equipment manufacturers provide these drawings.

Summary

Purchasing and warehousing provide a service to maintenance to ensure that required materials are made available, on time, and in proper quantities so that essential work can be done. In turn, maintenance must specify the materials needed, the quantity, the time needed, and how they will be charged. Within maintenance, shops manufacture spare parts or rebuild components to meet replenishment needs of the warehouse. Forecasting allows maintenance to anticipate material needs and provide purchasing with lead time to procure materials on time. Management guidance and policies should encourage this essential mutual support.

11

Training

How can supervisory training be improved?
What type of training must supplement skill training for craft personnel?

SUPERVISOR TRAINING

A Major Omission

Maintenance often fails to train first-line supervisors at all. Many good craft personnel are promoted to supervisor because of their personal technical repair skills. Too often they fail because they lack the management skills necessary to lead their crews effectively. Without proper training, they have little chance of success. Many general supervisors are promoted from the supervisor level, again without proper training to prepare them. Thus, poor supervisory patterns are merely transported to a new level where, by default, a higher order of disaster awaits! With a tendency to promote from within, superintendents are also often ill-prepared for the management challenges that they must meet. Relatively few have university degrees as compared with their operations counterparts. Many are, like their subordinates, graduates of the "school of hard knocks." Typically, they operate one or two levels below what their jobs require. For example, during a power failure, the superintendent was trouble shooting while well qualified electricians watched. A concurrent burst water line now found no one capable of directing action. Adding to this difficulty is the perception of management. Managers often see maintenance supervisors as the "folks who make the equipment run again!" The maintenance supervisor is seldom seen as the "manager" of his crew. The overall result of too little training and incorrect perceptions of the need for it is predictable. The maintenance program suffers. Typically, equipment inspections are missed, work is poorly organized, and proper work assignments are seldom given. Things just happen. Most are surprises, few actions are anticipated, prepared for, or controlled. If the organization is built around supervision that is not trained to perform its duties, they will have difficulty and the maintenance program will be poorly executed. Excessive cost and unnecessary downtime follow.

Need for Training

The first-line supervisors need for training is often apparent:

- There is little PM, emergencies prevail.
- Work is poorly controlled.
- Excessive cost and downtime are evident.
- Worker productivity is low.

At the general supervisor level:

- Planning is seldom adequate.
- Direction is missing.
- Accountability is absent.

At the superintendent level:

- The overall organization is unresponsive.
- The program is faltering.
- Equipment is unreliable with excessive downtime and cost.

Under such circumstances, training of supervision at all levels is required.

Objective of Supervisory Training

The objective of supervisory training is to improve maintenance performance with better control of work.

Thus, supervisory training should include:

Organization	Leadership
Supervisory techniques	Motivation
Work control	Counselling and discipline
Labor control	Work load measurement
Productivity	Training techniques

This should be followed by training that reinforces supervisor impact on work control, including:

Material control	Administrative procedures
PM techniques	Cost control/budgeting
Planning techniques	Performance measurement
Scheduling techniques	Maintenance engineering
Project planning/control	Terminology
Work order system	Productivity measurement
Information system	Shop operation
Use of the computer	Mobile equipment maintenance

The individual subjects should be defined and the content prescribed. Specific aspects of the educational areas would differ by supervisory level. For example, the maintenance superintendent would benefit in learning the theory behind preventive maintenance, its objectives, and benefits. The general supervisor would learn how to establish, operate, schedule, and verify compliance with the PM program. The first-line supervisor would improve his/her performance most by learning how he/she and his/her crew will execute the PM program. Each supervisory group should receive training on activities that will improve performance at their level. The first-line supervisor might, for example, receive instruction on the planning steps and techniques whereas the superintendent would be trained to develop a criteria for determining what type of work should be planned.

Conduct of Training

Discussions built on principles with interactions between the instructor and students are best. A seminar environment permits a "give and take" exchange in which individual questions can be asked and concerns satisfied. Supervisors must be treated as adults. They do not learn as children do; therefore they must have a rationale for learning and their egos must also be protected. Thus, caution is necessary not to put them back to school, create surprises, or present conflict with what they already know.

In conducting the training, consider that some supervisors may have limited reading skills and should not be assigned unnecessary reading. As the training is carried out, relate the materials to their jobs since they already know a lot but might be reluctant to ask questions. Create situations in which they can do something like setting up a daily schedule or creating a PM checklist. Conduct training in short sessions of an hour because supervisors are active personnel who do not normally sit down for prolonged periods.

Who Presents the Training

When every level of supervision is in need of training, get some outside help to avoid a "blind leading the blind" situation. However, once the training program is underway, push the senior supervisors ahead and gradually develop them as assistant instructors. As they conduct the instruction, listen to them explain material to junior supervisors as a means of verifying how well they understand it. This technique also builds confidence among senior supervisors as those who must, in the future, be able to sustain the long-term educational effort.

Check Progress

During training, the progress of individual supervisors can be gauged by interest shown, the quality of questions they ask, and the degree to which they participate. However, the real measure of progress is in the behavioral changes observed. For instance, a first-line supervisor who was indifferent about assigning work now provides daily work assignments for each crew member. A general supervisor who rarely attended weekly scheduling meetings now conducts them.

Benefits

As performance improves so will productivity and profitability. But, performance must be measured. Measurements must be directed at the following applicable supervisory levels:

1. *Supervisor:* Measure worker productivity, use of manpower, reduction in overtime, or compliance with the PM schedule.
2. *General Supervisor:* Performance improvement should be measured against cost and downtime reduction.
3. *Maintenance Superintendent:* Performance should be measured against manpower reduction, reduced shutdown time, or completion of a supervisory training program.

CRAFT TRAINING

Done Better

Craft training is conducted much more effectively than supervisory training. Unions have contractual arrangements requiring skill training. Skill training material is effective and readily available in a wide range of media. Manufacturers offer technical training. Craft personnel welcome and expect the training. Against this more encouraging picture, the most notable omission is training on the maintenance program. This omission is attributable to a failure of supervisors to be able to explain the program adequately to their crews. Not surprisingly, when supervisory training improves their performance, they train crews better.

OTHER TRAINING

Planners

Planners should be given the same training as supervisors and supervisors should be trained as planners. This assures that each will appreciate the support the other can provide.

Maintenance Engineers

Maintenance engineers should parallel the training of the superintendent. They should be aware of the specifics of supervisor and general supervisor training.

Plant Personnel

The mark of a well run plant is the degree to which management is familiar with the overall program, operations is conversant with program details, and staff departments understand how to support the program. Training of plant personnel on the maintenance program is an essential requirement in their support of maintenance.

Summary

Supervisory training is a neglected area requiring attention. However, when it is carried out effectively, gradual improvement can be noted. But, performance measurements must be made in order to check progress. Craft training is usually well done when only skill training is involved. Unfortunately, training on the maintenance program is often omitted. As a result, craft personnel who could contribute have no framework for helping. Poor supervisory training and the absence of program training for craft personnel are related. It is the maintenance supervisors who must conduct such training. If they do not understand the program, they cannot explain it to their crews. Not surprisingly, when the void in supervisory training is filled so is the need for program training for craft personnel. Supervisory training is an absolute requirement in improving maintenance performance. Therefore, when overall maintenance performance is inadequate, look first at the quality of the supervisory training effort. It follows that any manager who desires improved maintenance performance will do well to start with the education of maintenance supervision.

II

What Should You Examine to Establish That:

The plant environment is conducive to the success of maintenance?

Maintenance is organized to respond quickly and effectively to its objectives while making the best use of its personnel?

The maintenance program is well-defined and understood by those who carry it out, use its services (operations), or support it (staff departments)?

Maintenance performance is assessed and steps are being taken to achieve improvement?

12

Environment: Encourages Success

How can you create a supportive, cooperative plant environment that will ensure the success of the maintenance program?

ESTABLISHING A FAVORABLE ENVIRONMENT FOR MAINTENANCE

Management Support

As manager you must establish the overall production strategy—the plan for achieving profitability—and assign key departments mutually supporting objectives within that strategy. The objective you assign to maintenance must emphasize the principle roles you expect them to perform to help carry out your strategy. In turn, you should amplify the objective with policy guidelines (ground rules) that spell out the desired interaction between key departments and maintenance. Based on these policy guidelines, maintenance would develop procedures and establish a concept of how work should be requested, planned, scheduled, assigned, controlled, and measured. Overall, you must create an environment in which the maintenance program can be carried out effectively. You should verify that you have:

1. Assigned a specific objective to maintenance.
2. Provided policies emphasizing maintenance procedures that have a plant-wide impact.

3. Reviewed the overall maintenance program.
4. Specified how maintenance performance will be measured and assure that it is measured.
5. Taken into account maintenance work force limitations such as their capability to perform construction work.

Production Support

Maintenance is a service, the primary objective of which is to keep production equipment in a safe, effective operating condition so that production targets can be met on time and at least cost. Thus, production, as the principal benefactor of this service, must understand the service, cooperate, and follow up to ensure that cost and work quality are satisfactory. You should determine whether:

1. Production personnel display a comprehensive knowledge of the maintenance requirements of the equipment that they operate.
2. The maintenance program is facilitated by prompt release of equipment for repair.
3. There is minimum blockage or damage due to mal-operation.
4. Production participates in the development of the maintenance program (like PM services) and reviews its accomplishments (like cost reduction).
5. Production is involved in the identification, timing, and cost of future major work like overhauls.
6. Compliance with PM services and timely completion of major jobs are checked by production.
7. Operators report problems promptly.
8. Operators are effective in carrying out cleaning, servicing, and simple adjustments of equipment.
9. Production is concerned with job cost and quality of work.
10. Production does not overload maintenance with unexpected work or unreasonable demands.
11. All major work requested by production is necessary, feasible, and correctly funded.

As a result of these actions, cooperation between maintenance and production should be evident at every level of their respective organizations.

Objective

A maintenance department must have a clear statement of its objective to ensure that its activities are consistent with the overall plant production strategy and the organization's total objective. The maintenance objective is usually accompanied by mutually supporting objectives for other departments. Typically, the primary objective of maintenance is to maintain equipment, as designed, in a safe, effective operating condition to ensure that production targets are met both economically and on time. Maintenance will also support non-maintenance project work (such as construction) as required. In addition, maintenance will maintain buildings and facilities and provide support services such as boiler operation or power generation. You should determine that:

1. There is a maintenance objective and it is stated clearly.
2. The objective describes the important functions to be performed in support of your plant production strategy.
3. The maintenance objective has been effectively communicated to the total plant organization.
4. The clarity of the objective has eliminated any confusion about what maintenance should be doing and the relative priorities of its actions.

Policy Guidelines

Maintenance is a service organization and it cannot compel production to comply with its program. Therefore, you should utilize policy guidelines to emphasize important maintenance concepts which have impact throughout the organization. This assures that these concepts are called to everyone's attention. Policies are also issued to ensure that important maintenance procedures are followed with reasonable uniformity on a plant-wide basis. In a typical policy, maintenance will develop and carry out a preventive maintenance program to inspect, test, lubricate, adjust, and clean equipment to avoid premature failures and extend equipment life. Also, production will cooperate in the preventive maintenance program and determine that prescribed services were carried out effectively as scheduled. You should also verify that policies emphasize adherence to important maintenance concepts, important procedures are backed up with firm management policies, and maintenance procedures are followed.

Plant Staff Support

Staff support includes activities such as warehousing, purchasing, data processing, and so forth. The support of the maintenance program by staff departments and their personnel constitutes a service without which maintenance could not carry out its program effectively. Staff personnel must understand the maintenance program, provide procedures for the use of their services, and render a high level of service to maintenance. To assess staff support, you should determine whether:

1. The typical staff department member has made a genuine effort to understand the maintenance program.
2. Staff personnel are service-oriented.
3. Recommendations for the modification of administrative procedures used by staff departments are considered and acted on.
4. Staff departments confer regularly with maintenance to improve service.

Material Control

Material control includes procedures to identify materials, stock, purchase, or manufacture them in proper quantities to make them available to maintenance as requested. To obtain materials, maintenance must determine what it needs and specify the quantity and the time when they are required. The work order system, material control, and information systems permit maintenance to obtain information on material costs and usage. Effective material control is sufficiently important to maintenance for you to verify that:

1. Material control procedures covering stock, purchased materials and "in-house" manufactured items are well explained and effective.
2. Critical spare parts lists exist for important equipment.
3. Maintenance specifies the repair materials it wishes stocked and the initial levels.

4. When materials needed are out of stock, prompt actions are taken to replenish them.
5. The stock room is properly staffed.
6. The procedure for withdrawing stock material is well organized.
7. A procedure exists for the return and crediting of unused stock parts.
8. Accountability for stocked parts is effective.
9. Maintenance personnel follow prescribed stock withdrawal procedures. (They are not allowed, for example, to "prowl" through the stockroom in search of parts.)
10. Stock material can be identified by a company number as well as the manufacturer's number.
11. Parts interchangeability between manufacturers is well documented.
12. Reference materials are available to identify materials with correct part numbers and descriptions.
13. Maintenance craft personnel can identify parts properly.
14. Multiple parts can be listed on a single stock withdrawal card.
15. A procedure exists for reserving stock parts in advance.
16. Standard bills of materials for major repetitive jobs such as overhauls or rebuilds exist.
17. Stock issues can be made directly to craftsmen who present approved work orders without further approval by maintenance supervisors.
18. The stock withdrawal document cross-references parts and quantities with either work orders or equipment numbers.
19. The cost of material consumed in the repair of a unit of equipment is summarized each month.
20. The cost of stock and purchased materials consumed on a work order can be obtained job by job.
21. Lead time required for purchased materials is carefully observed by maintenance.
22. There is a purchase order tracking system.
23. Drawings necessary for local manufacture of spare parts are available.
24. An efficient on-site material delivery system exists.
25. There are no "bootleg" (unauthorized) material storage areas.

These actions should ensure that overall material control is excellent.

Summary

As a service organization, maintenance can neither compel operations to comply with its program nor can they demand services from the warehouse, for example. Yet, this cooperation and support is as essential to maintenance as is their contribution to plant profitability. Your assignment of a clear objective and policies to guide department interaction will help to ensure cooperation from operations and support from staff departments. Further, you should insist that maintenance define its program and educate personnel on its provisions. Material control is seldom controlled by maintenance, but it is a key factor in their success. Thus, it may require your attention to assure its effective operation and the quality of support for maintenance. Collectively, these matters influence the quality of maintenance performance. Your attention to them can create a positive, supportive environment for the success of maintenance.

13

Organization: Responsive and Efficient

How can you determine whether maintenance is organized properly to meet its assigned objective?
How can you ensure that they have placed personnel in the best position to use their talents effectively and obtain quality performance?
How can you measure organizational performance and convert needs into improvement actions?

ORGANIZATION

Traditional Organization

When there is a supervisor hierarchy, correct, complete organizational charts should exist and show the chain of command on all shifts. The organizational structure should encourage working efficiency while facilitating communication and interaction. Job descriptions should be up-to-date and describe actual duties. During peak vacation periods, there should be sufficient depth and flexibility to still get work done.

Personnel

Criteria should exist for the selection of key personnel. Supervisors and crew members should be assigned to production areas or shops based on the actual work required. First-line supervisors' spans of control should permit effective control of crews. There should be enough supervisors to provide vacation or absentee coverage. Their duties should allow them time to supervise—their key duty. Supervisors should be able to control the work of several different crafts well and craft jurisdiction should not be a problem. Crews should be able to respond quickly to work requests and make decisions automatically on actions they control.

Control of Activities

Procedures for work-control should be documented and understood so that personnel can respond promptly to work requests. Control functions like preventive maintenance and planning or non-maintenance work (like construction) should be carried out efficiently.

Team Organization

The reality of smaller, more productive maintenance work forces with "self-directed teams" achieving prodigious amounts of quality work at vastly reduced costs has eluded many organizations. Urged on by managers who visualize astounding Japanese-like achievements, maintenance organizations are struggling mightily to emulate them—often unsuccessfully and with considerable frustration. Advisers, consultants, and corporate specialists abound and eagerly offer their services on how to do it. They will:

1. Train discussion facilitators.
2. Show how to conduct meetings.
3. Demonstrate how to get captive groups to list "interesting suggestions" like painting stairs yellow and expanding lunch rooms.

Unfortunately, too few such advisors understand the industry or the maintenance environment within which the advice is offered. As a result, very few substantive suggestions emerge on the one topic it is all about— improving worker productivity and performance. Somehow, the promise of smaller maintenance work forces operating without supervision and working efficiently to yield high quality work at lower cost has not materialized in many instances.

IMPLEMENTING MAINTENANCE TEAMS

A Ten Point Program

Here are ten points to consider in implementing maintenance teams:

Point One: A Team Probably Will Not Succeed Unless Employees Think That They Need It As Badly As Management Wants It. The Japanese have succeeded with the team concept, in part, because they are the homogeneous society that we are not. For example, try visualizing a crew of burly mechanics doing their morning calisthenics with their supervisor or explaining a union picket line to visiting Japanese businessmen. Better quality work is not necessarily an exciting enough objective until it threatens plant closure. Then teams are possible. This also explains why an unprofitable plant sold to its employees is often made profitable in record time.

Point Two: Level with Employees. Employee productivity, improved performance, and cost reduction are among the expectations of management for the maintenance team organization. However, management also knows that the cost of maintenance cannot be reduced unless it is done with fewer people or less often. Employees also know these facts. Therefore, a credible explanation must be offered on why a team is thought necessary. If the maintenance work force must get smaller to reduce costs, then preserve some jobs by displacing outside contractors with current maintenance employees.

Point Three: Straighten Out the Maintenance Program. The principal cause of the failure to successfully implement a team organization is the absence of a well-defined maintenance program within which the team is expected to operate. A self-directed team operates well if they are presented with well-identified, ready to go jobs at the start of each shift. Their strength is in sorting out who does what and in what order. Conversely, if the team is bombarded with requests for emergency repairs, it responds to the most life threatening and then retreats to the sanctuary of their "lunch room shop" and lets the dust settle. While many team members appreciate that continuous equipment condition monitoring can prevent emergencies, it is simply not their responsibility to create the monitoring program. Plant engineering must create a viable program into which the efforts of the team fit. Organizations aspiring to implement a team successfully must bring the team to life within the framework of a well-defined maintenance program.

Point Four: Do Not Forget the Traditional Supervisor. Conversion of the traditional organization to a team is more difficult than starting new. The biggest surprise is resistance from the maintenance supervisor—presumably part of management. He sees himself becoming obsolete. Because his future is threatened and uncertain, he fights to preserve the status quo or delay the questionable future that faces him. When converting a traditional organization, assure the supervisor of his future task in the team organization.

Point Five: Recruit the New Team Carefully. Generally, organizations that recruit directly into a new nonunion organization stand a better chance of success. Their personnel directors are able to spot and reject the "storm trooper" type in favor of the potential team player and, as a result, the team starts out with fewer handicaps. Despite this, many unknowing personnel managers hire "frustrated contractors" thinking that they are getting maintenance craft personnel. Soon, essential preventive maintenance services that help guarantee reliable equipment are displaced with questionable projects encouraged or even solicited by these folks. Recruiters should look for personnel with good craft skills who exhibit the potential for team membership. However, to these qualifications must be added:

1. An appreciation of how the "total" maintenance program must be carried out
2. How it fits within the manager's production strategy (his plan for using maintenance services to achieve profitability)

If prospective team members appreciate the framework of maintenance management, their contribution to a successful maintenance program is much better. By starting with the right type of people, the plant can avoid taking on difficult behavioral changes that are necessary to make the team functional.

Point Six: Decide on the Salaried Status Carefully. The maintenance team will be made up of former hourly workers whose status may now be changed to that of salaried employees. These new salaried employees may be looking for the "perks" (like flexible lunch periods) that other salaried employees seem to enjoy. They must realize that their work status and shift assignments simply do not permit them the same latitude afforded to other salaried personnel. Equipment maintenance needs do not recognize individual preferences. When the need arises, there must be a rapid, dependable response. To help them make the transition, provide ground rules to clarify their expectations. Ensure that personnel know that:

1. Equipment maintenance requirements set their work schedules.
2. Vacation periods or lunch hours may lack the flexibility of those provided to office workers, for example.
3. As workers as well as decision-makers, they may have constraints on what decisions they can make. They can specify the order in which they do work providing that it is coordinated with operations, but they cannot decide whether or not to do certain work requested by operations.

Point Seven: Establish Team Decision-Making Parameters Early. A self-directed team is qualified to make decisions on how to perform work. However, if they, for example, are asked to decide how to manage maintenance or which shifts they will work, it may ask for more than they can deliver. Neither the decision nor the result will be satisfactory. Determine in advance the boundaries within which the teams can make decisions. The team can schedule established PM services after coordinating with production, but they cannot specify what services are to be performed. However, they could recommend changes to existing programs. Also, the team can modify equipment only if prescribed by plant or maintenance engineering. But, they can recommend how the modifications might be made.

Point Eight: Determine How Craft Skills Will Be Evaluated and Remedial Training Provided to Improve Lagging Skills. A team can become a hiding place for the incompetent worker who needs skill training. It is unlikely that he/she will be exposed and turned out by his/her team. Consequently, the desired improvement in work quality may never materialize. Provide continuous skill training, but also establish procedures for evaluation of skills and provision of remedial training when it is found to be necessary.

Point Nine: Give the Team Firm Work Control Procedures, As Most Team Members Have Not Had the Training or the Experience to Develop Their Own. Within a framework of reasonable work control procedures such as daily work assignments and determining which type of work requires planning, most teams can be successful. For example, at a steel pellet plant, maintenance personnel on back shifts operated without maintenance supervision. This arrangement thrust craft team members into a situation requiring that they call "their own shots." Fortunately, they had received considerable training on the overall maintenance program and, as a result, saw themselves as the direct maintenance contacts on the back shifts. With this education, they saw the total picture of work control and were able to competently "manage" work requests, unscheduled repairs, and handle "surprises." Often, they did this with praise and compliments from their production counterparts. Thus, education on the maintenance program framework within which the "team" effort must fit is essential to their success.

Point Ten: Organize Maintenance and Establish a Working Relationship with Operations That Fosters Team Utilization and Development. For example, in an underground coal mine, each 8-man production crew was assigned two mechanics. The mechanics realized that their success would depend on working as team members with the operators on the crew. They not only made it clear to operators what services they would perform, but when they must be done. To solidify the team working relationship, mechanics tried to be helpful to operators. If a continuous miner operator needed a break, for example, a mechanic relieved him. When the operator returned to the unit, the mechanic provided pointers on checking the unit during operation. In turn, if there were heavy repair jobs to be done, the mechanic could count on the operators for help. No one told them to do it, they just did it. This type of maintenance-operations cooperation made the team successful. Self-directed teams can be beneficial to both the organization as well as to team members. The team will become successful more quickly if it is developed within the framework of a well-identified maintenance program. Sensible guidelines on team decision making and skill training should be applied as the team is being developed rather than after it is discovered that these are problem areas. Conversion of traditional organizations will be more successful when the supervisor, who is about to be displaced, is given an honest picture of their future role. Most important to the success of a team is the creation of an environment in which the team can succeed including solid reasons for employees to want the team as much as management does.

HOW MANY PEOPLE

Work Load Measurement

This is the identification of the essential work to be performed by maintenance converted into a work force of the proper size and craft composition. Once determined, the work force must be able to be adjusted as work loads change. Special care is necessary to ensure that non-maintenance work (like construction) does not interfere with the basic maintenance work load. Any work force increases should require justification before they are approved.

SUPERVISION

Selection and Promotion

A criteria should exist to ensure that the best candidates are chosen as supervisors. Selection of senior maintenance supervisors should not be restricted to maintenance personnel as talent from production or engineering can enrich maintenance. A clear line of promotion should exist for supervisors and promotion criteria prescribed to reach each higher level. The position of supervisor should be respected enough that qualified craft personnel seek it eagerly. Conversely, if supervisors promoted from craft ranks seek return to their tools, there can be serious dissatisfaction.

Duties and Responsibilities

Ultimately, the supervisor is responsible for everything the crew does or fails to do. Therefore, supervisors should have specific responsibilities and sufficient authority to carry them out. To ensure that they have adequate time and opportunity to carry out their key duties, they should have minimum administrative tasks assigned.

Performance

Supervisors should be rated on their performance and advised of the results. Those who perform poorly should be counselled to help them develop more effectively. If found incompetent, they should be replaced promptly. Maintenance supervisors must understand the maintenance program so that they can use their crews to carry the program out effectively. While they must be technically competent themselves, they should be confident in delegating responsibility to crew members.

Training

There should be an effective maintenance supervisor training program solidly backed by management. Since the task of controlling the maintenance work is in the hands of maintenance supervisors, they must be prepared for this task through training. Then, to continue to be effective, they must keep their supervisory skills up-to-date. Supervisors are also required to train subordinates, whether they be other supervisors, planners, or craft personnel. Thus, their communication skills must be considered in their training along with other essential maintenance management techniques. Newly selected supervisors should be trained before they assume their duties. Similiarly, when supervisors are identified for promotion, there should be training to help them prepare for the broader duties and responsibilities that they will assume.

Supervisors should receive training in the maintenance management program so that they can competently carry it out and communicate it effectively to their subordinates. Work related matters such as labor control, work control, and cost management should be emphasized. Training in human-relation aspects of supervision such as leadership, motivation, counselling, and discipline would be helpful. Supervisors should also be trained in the duties of maintenance planners to appreciate the value of planning and, if required, temporarily replace a planner. They should receive training in production processes and their hazards. Supervisors' capabilities and job interests should be enhanced by job rotation and off-the-job training such as seminars when possible. A procedure should exist for determining supervisory training needs with a method for converting needs into training. Up-to-date training records are necessary to provide a complete picture of training received and progress achieved.

CRAFT PERSONNEL

Training

There should be an overall craft training program with strong management backing. Craft skills must keep pace with current technology. Basic skills must be kept up-to-date. Training materials and methods must be effective. Those giving the training must be well qualified. Training should challenge and motivate those being trained. Personnel who are to be trained should meet certain minimum standards to qualify for training.

Testing should be carried out to ensure that training is effective and that those trained did learn. Attendance and progress records should be kept. Progression from one level to the next should be conditional upon demonstrating necessary skills. Higher skill levels should never be awarded on the basis of seniority. When greater skill levels are achieved, personnel who qualify should be acknowledged. New maintenance employees should be properly indoctrinated immediately on joining the work force.

Program Checkpoints

The craft training program should provide for the selection and training of personnel to fill craft needs. Technical training for craft personnel should be carried out on a regular basis. Training on the maintenance program should be mandatory for all maintenance personnel. On-the-job training of craft personnel by supervisors should be augmented with classroom training. Specialized training sessions should be conducted on specific maintenance problems, particularly those requiring unusual corrective measures. Training by equipment manufacturers should be included in the program as required. Training should ensure that craft personnel become skilled in diagnosis and repair and that they are not merely "parts changers." Craft personnel, including contractors, should be trained in the basic production process and related hazards.

Training Administration

Official records should be kept on training received and progress achieved. Suggestions for improvement in maintenance procedures by craft personnel should be solicited, reviewed, and, when warranted, converted into training sessions. Training techniques, such as video or programmed instruction, should be carefully matched to training needs. Higher skill levels should not be awarded on the basis of seniority. All personnel must successfully complete required training.

Results

The training program for craft personnel should result in tangible benefits such as reduction in the need for contractors, less rework, and fewer overtime hours.

MOTIVATION

Motivation

Motivation is a stimulus to action. When applied properly, motivation results in improved productivity, better quality work, and greater job satisfaction. Individuals are motivated when they feel that their jobs are important, their roles in carrying out those jobs are recognized, and their rewards for performing those jobs are regarded at the same level at which they see it. Motivation applies to all individuals—from manager to laborer. While the form of reward at different organizational levels may differ, the essentials of motivation at each level must be met. An otherwise excellent maintenance program will not be successful unless its work force is motivated to make it successful.

The maintenance manager should be able to motivate the maintenance work force. In turn, supervisors should motivate their crews and they, themselves, should be well motivated as should staff personnel, such as planners. It follows that if the hourly work force is well led, they will be motivated and satisfied with their role, degree of recognition, and level of reward for the work they do.

Summary

Whether maintenance organizes along traditional lines or utilizes a team, the objective of the organization is to place personnel in the best posture to execute the maintenance program effectively. Selection, training, and promotion must be carried out so that personnel know what is expected of them, and that they are prepared to do it properly. They should be assured that recognition and promotion for quality performance are provided for.

14

Program: Understood and Supported

How can you ensure that the maintenance program is well defined?
How can you determine whether it is understood by the maintenance
personnel who must carry it out?
How will you find out if operations understands how to utilize maintenance program services effectively?
How can you ascertain whether staff personnel, such as warehouse supervisors, appreciate the importance of their support for the maintenance program?

DEFINING THE MAINTENANCE PROGRAM

Concept

The failure of many maintenance programs can be traced to inadequate
definition and a failure to educate those who must carry them out and
use their services. If your aspirations are an effective team organization, keep in mind that most that fail do so because they were introduced into an environment in which the program they were expected
to perform was ill-defined.

The adequacy of program definition can be determined by examining its conceptualization—the way maintenance services are handled
from request to completion. While this appears to be a major task, it can
be divided into controllable segments. However, it must be verified be-

cause, if effective maintenance is the objective, it starts here. See Figure 2-1. You should ensure that:

1. There is a concept of maintenance and it has been described clearly, using, as necessary, schematic illustrations, standard operating procedures, and so forth.
2. It describes how maintenance services are requested, planned, scheduled, assigned, controlled, carried out and, on completion, how cost and performance are measured.
3. It has been communicated to all personnel and they understand how to carry it out (maintenance), use its services (operations), and support it (staff departments).

The concept explanation will use terms possibly unique to maintenance. Therefore, it must be accompanied by a definition of the terminology used.

THE LANGUAGE OF MAINTENANCE

Terminology

Terminology is the language of maintenance. You should ensure that it includes terms and definitions by which maintenance and maintenance personnel and their customers and supporters can communicate effectively. See Appendices D and E.

MAINTENANCE COMMUNICATIONS

Work Order System

The work order system is the maintenance communications system. It is the means by which work is requested, planned, scheduled, and controlled. It focuses labor and material data from accounting documents like the time card or stock issue card so that it can be processed into information. To establish the adequacy of the work order system, you should verify that:

1. The work order system (including provision for verbal orders) covers all types of work.

2. It is used for requesting, planning, scheduling, and controlling all work done by maintenance.

3. Operations has been provided with a simple way to request work to avoid unnecessary verbal orders except for emergency work.

4. Verbal orders are an effective part of the system but are used only for emergency work and, when used, get work done with no loss of control or information.

5. It contains provisions for job planning and estimating while permitting selected major jobs to be measured against cost and job performance targets.

6. It controls routine, repetitive activities like grass-cutting, janitorial work, or safety meetings in the simplest possible way.

7. Standing work orders are used only for routine, repetitive functions like preventive maintenance or grass-cutting, or to represent large groups of low maintenance cost equipment in the same cost-center, none of which warrants cost management alone.

8. Standing work orders are not substituted for equipment numbers. (This is a sure sign of an inadequate work order system and indifference on the part of maintenance about cost control.)

9. It provides for the control of non-maintenance jobs like construction and makes provision for controlling contract work which cannot be supported by standard accounting documents. (The contractor will not accrue labor cost with the company time card. Rather, the contractor will invoice labor costs derived from accounting procedures.)

10. It is effectively linked to time cards, stock issue cards, and purchasing documents and tied into the information system to produce useful management information.

11. Work order documents are to be simple and easy to use and permit small jobs to be handled in a simple way without loss of control.

12. Instructions on how to prepare, use, and process work order documents are complete and clear.

13. Priority-setting procedures are practical, realistic, and followed.

14. Major work is approved by equipment users and criteria exist for approval based on cost and need.

Overall, these verification steps ensure that the work order system is used properly, resulting in solid work control, quality information, and effective management.

PREVENTIVE MAINTENANCE

Objectives

The Preventive Maintenance (PM) program should successfully avoid premature failures through timely inspection and testing. It should be "detection oriented" to find problems before they create equipment failure. Predictive maintenance, also called nondestructive testing, combines with physical equipment inspection to comprise "condition monitoring." In addition, PM should extend equipment life with lubrication, cleaning, adjustment, and minor component replacements. As the result of the "detection orientation" of PM, there should be fewer emergency jobs and more work should be able to be planned because deficiencies are found soon enough to plan the resulting work. As the planned work is performed, maintenance personnel will work more productively and deliberately. Thus, the results will have lasting quality.

Education

The program should be explained to operating personnel to enable them to cooperate and use its services effectively. Operators should perform routine PM-related tasks like lubrication, cleaning, and adjusting to help ensure dependable operation of equipment. A surprising number of maintenance personnel in the same organization define PM differently. An equally surprising number of operations personnel have had no education on its content, purpose, and the role operators must play in supporting it. Unfortunately, there are still too many managers who think that PM is the whole maintenance program and sadly, too few bother to correct this misunderstanding. To ensure that the preventive maintenance program effectively supports your plant operations, you should verify that:

1. For starters, everyone understands what preventive maintenance is and how it is carried out.
2. Each PM service has a standardized checklist which explains how and when the service is to be performed.
3. Service frequencies are correct and a procedure exists to adjust them as required.

4. The manpower required for each PM service and for the entire PM program has been determined in advance.
5. Procedures exist to verify the quality and adherence to the service schedule.
6. Nondestructive testing actions such as vibration-analysis or infra-red testing have been integrated into the PM program and determination made whether they will be contracted or done in house.
7. The program is reviewed periodically and updated to reflect changing conditions.
8. New equipment is added to the PM program promptly.
9. During the conduct of inspections, extensive repair actions are avoided since they too often divert the inspector from completing the inspection. Thus, problems more serious than those corrected go undetected.
10. PM services requiring equipment shutdown are scheduled in advance to avoid unnecessary interruption of operations.

Through these actions, you will cause PM to successfully reduce emergency work, extend equipment life, and enable maintenance to plan more work and assure more productive use of its work force.

INFORMATION

Categories of Information

Maintenance information falls into two categories:

1. Decision-making information to manage the maintenance function.
2. Administrative information to provide clerical information on such matters as absenteeism, use of overtime, open or closed work orders, and so forth.

Objectives of Information

Information provides the basis for control of work, use of resources, cost performance, and so forth. Information must then be effectively communicated to the lowest level consistent with the need to control activities effectively. In your assessment of the maintenance information system, you should determine that:

1. Essential information to manage maintenance is readily available and content and use are understood.
2. Indices to gauge overall maintenance performance are provided to quickly spot trends in cost and performance.
3. Production actively reviews maintenance cost and performance.
4. All personnel who need information are able to obtain it and they are proficient in its use.
5. Labor utilization information should show the amount of labor used on PM, emergency repairs, scheduled maintenance, and so forth; information on absenteeism and overtime use are available and informative.
6. Backlog information shows how well maintenance is keeping up with the generation of new work on a craft by craft basis.
7. There is an open work order list which shows all work orders currently open or being worked on unit by unit.
8. Information is available on the cost and performance of major jobs from start to finish and, on job completion, cost and performance are summarized.
9. Actual and estimated man-hours is able to be compared for selected major jobs.
10. Cost information is summarized (month and year to date) for units of equipment (like conveyors) and components (like drive motors) by cost-center.
11. Cost information is provided on routine functions like grass-cutting or janitorial work.
12. Summarized actual maintenance costs for each cost-center are compared with budgeted costs on a month and year to date basis.
13. Repair history traces significant repairs and equipment failure patterns on key units of equipment and includes information on the life span of critical components to aid in predicting their future replacement.
14. Report formats or computer screen displays are informative, clear, and easy to read.
15. The information system satisfies genuine information needs and it is not, for example, a hastily purchased package of limited practical value or an awkward adaptation of an accounting report.

These verification steps should result in quality information that is properly applied to result in the effective management of maintenance.

PLANNING

Planning Definition

Maintenance planning is the administrative preparation of selected major jobs in advance so that upon execution they can be completed more efficiently.

Objectives

Planning determines resource requirements such as labor, materials, tools, and supporting equipment (like mobile cranes) to assure their availability for specific jobs. Planned jobs use less manpower, are completed with less downtime, and the work is done more effectively. Thus, the repair is of lasting quality. There should be criteria for the selection of those jobs which will be planned. There must be an effective procedure for planning. Planners should be used primarily for planning work. In verifying the quality and effectiveness of maintenance planning you should determine that:

1. A criteria specifies conditions under which jobs are to be planned and assures that important major maintenance work is planned.
2. The planning procedure is complete and understood.
3. Work order documents support the details of planning and provide information on the cost and performance status of planned jobs.
4. Repair history data on the life span of major components is used to develop a component replacement program (periodic maintenance).
5. A long-range forecast identifies future periodic major jobs requiring planning and scheduling.
6. PM inspections are a major source of planned and scheduled work.
7. Planned work is developed from failure analysis and nondestructive testing results such as oil sampling or vibration-analysis.
8. Major planned jobs are approved by equipment users before work is done and approval is based on estimated cost and need.
9. The priority system prescribes the relative importance of a job and the time within which it should be completed.
10. All planned work is prioritized to assess its comparative importance and facilitate allocation of resources.

11. All planned work is scheduled and labor is allocated according to priority.
12. A criteria exists for the selection of enough planners to do required planning.
13. Planners are well trained and have ample time to plan.
14. Key operations personnel understand the value of planning and insist maximum work be planned and scheduled.

By completing the suggested checks, you can ensure that planning will successfully reduce both the manpower and elapsed downtime required to complete major jobs.

SCHEDULING

Objectives

Scheduling of major jobs should be a joint maintenance-production activity in that maintenance provides the resources to do the work while production makes the equipment available. The two actions must coincide to establish the best timing for a job to avoid interruptions to production and utilize maintenance resources most efficiently. Repetitive tasks such as PM services are scheduled routinely by maintenance after having informed operations of the master schedule on which weekly incremental scheduling is based. To determine the effectiveness of joint maintenance-production scheduling, you should establish that:

1. A policy exists requiring the joint scheduling of activities impacting the harmonious interaction of operations and maintenance.
2. A weekly schedule is presented by maintenance supervisors at a joint meeting and operations approves it before work is done.
3. Once approved, operations makes equipment available according to the schedule.
4. On a daily basis, operations and maintenance should coordinate or adjust work from the weekly schedule applicable that day.
5. Manpower for scheduled maintenance and non-maintenance work like equipment modification is allocated by priority.
6. Material needs are anticipated and then delivered by material control personnel so that shortages rarely interfere with the schedule.
7. Attendance at weekly scheduling meetings includes the right maintenance and operations decision makers.

8. Operations does not override an approved schedule without solid justification.
9. Man-hours spent on scheduled work are measured by major job as well as by weekly work load increments. (This assures that PM inspections are creating the opportunity to plan adequately.)
10. Compliance is measured on the completion of each week's schedule showing the number of jobs successfully completed versus those scheduled.
11. Work that is planned and scheduled is completed within the time prescribed by the priority assigned to it.
12. Based on the approved weekly schedule of major jobs, the PM service schedule, and the opportunity for equipment shutdown, each maintenance supervisor prepares a daily work plan to guide the assignment of jobs to individual crew members.
13. Individual maintenance supervisors or team leaders have a work assignment and control procedure that facilitates getting work done on time.

By following up on the suggested checks, you will assure that interdepartment cooperation is enhanced and that elapsed downtime required to perform major work is minimized.

USING STANDARDS

Definition

Standards provide targets against which maintenance efficiency can be evaluated. Generally, there are two types of standards.

1. Engineered standards that are applied mostly to unique major jobs specify manpower use, job duration, job tasks, and the quality level of the finished job.
2. Historical standards that are applied to repetitive tasks such as the periodic replacement of major components like engines or drive motors.

Maintenance planners should take advantage of current computer technology which summarizes and stores historical data on man-hours used or materials consumed. Such data can verify job performance and help create numerous additional historical standards. As a result, the task of the planner is simplified and, in some cases, the number of planners needed is reduced or planning is carried out by supervisors or teams. You should verify that:

1. Standards are used for jobs on which work quality, cost, and performance are important.
2. The work order system supports the use of standards by comparing actual performance with standards.
3. The information system provides feedback on job performance which can be used to develop or confirm historical standards.
4. A procedure exists for the use, preparation, and updating of standards.

In making these checks, you can assure that the best use is made of engineered standards to extract good performance while historical standards simplify planning and improve its quality.

BACKLOG

Definition

The backlog measures the degree to which maintenance keeps up with the generation of new work while helping to adjust the size and craft composition of the work force as the work load changes. Be aware that the backlog to some is simply a list of incomplete jobs. Thus, an increasing list of big jobs, little jobs, emergencies (overlooked), and planned jobs all mixed together may signal that maintenance is falling behind. However, it is useless in adjusting the work force size and craft composition. To accomplish this, the backlog must reflect estimated craft man-hours by job and type of work. Thus, your assessment should determine that:

1. Maintenance understands what the backlog means.
2. They have a current picture of the changing work load.
3. They are aware that the size and composition of the work force should change if there are significant, continuing differences in the backlog level.

By determining the quality and understanding of the backlog, you can assure yourself of a means of controlling the size and composition of the maintenance work force.

SUPPORT SERVICES

Definition

Transportation to move people and materials, rigging, the provision of support equipment such as mobile cranes, and the operation of facilities such as power generation, hoist operation, heating, or compressed air are types of support services maintenance provides. In addition, buildings and grounds work, custodial services, and road and parking area maintenance are support services commonly carried out by many maintenance departments. To assess the adequacy of support services provided by maintenance you should determine whether:

1. Transportation and support equipment can be obtained readily.
2. Rigging for large field projects is carefully planned.
3. Personnel are qualified to operate support equipment.
4. The cost of operating equipment is charged against the job on which it is used.
5. There is a sufficient amount and variety of equipment.
6. Custodial services or buildings and grounds maintenance are carried out effectively.
7. Scrap material, trash, and debris are properly collected, classified, and disposed of.
8. Service roads are well maintained. In winter, where applicable, snow removal is carried out effectively.

By assessing support services, you will be able to ease the conflict often found when operations and maintenance compete for the use of equipment and services provided.

SHOP OPERATION

Definition

Shop facilities such as a machine shop, fabrication or welding shop, carpentry shop, paint shop, and plumbing shop, provide support for field maintenance work as well as non-maintenance project work like new installations or construction. In your review, you should verify that:

1. A policy exists requiring that shop customers provide clear instructions and, in failing to do so, the shop may refuse to perform the work.

2. Shop operations such as lathe work, welding areas, assembly points, or finished work storage are carefully laid out to ensure smooth, efficient flow of work through the shop.
3. Shop work stations (such as a lathe or an assembly area) are designated and work planning is carefully related to them to ensure effective work control.
4. The work order system depicts the sequence of steps necessary to control work.
5. Information is available on the status of major shop jobs.
6. Emergency work does not reduce shop operating effectiveness.
7. Drawings, schematics, or diagrams necessary to carry out shop work are complete and well organized.
8. Procedures for requesting shop work are well explained.
9. Shop performance and productivity are measured.
10. A priority system exists which allows the shop to allocate manpower, machine time, or work areas to satisfy both maintenance and engineering customers.
11. Shop personnel are able to be assigned to field work with no loss of shop efficiency.
12. Shop tools, dies, or materials such as shop stock, sheet metal, bolts, and welding supplies, are properly stocked and well organized.
13. The consumption of materials, use of manpower, and machine time are brought together by the work order and information system to show job costs.
14. Tool control is carried out effectively.

As a result of your assessment, shop personnel should be able to carry out their jobs skillfully while supervisors and planners control shop work.

NON-MAINTENANCE WORK

Definition

Maintenance departments are often called upon to perform non-maintenance work such as construction, equipment installation, and modification. The same craft skills are generally used but, in some instances, a separate element of maintenance may perform these activities. In many maintenance departments, three potentially troublesome problems persist:

1. Operations may insist on equipment modification without first establishing the necessity and feasibility of changes that they feel they need.
2. Too often, these changes are not subjected to either engineering scrutiny or funding criteria. Thus, if completed, the work may undermine the process flow as well as distort the maintenance budget.
3. Some maintenance personnel are "frustrated contractors" who actively solicit such work because they prefer it to the troublesome (for them) task of diagnosis and repair under sometimes unfavorable conditions.

A major problem can be averted by establishing a policy on how much of this work is to be performed and the methods by which it is approved. Your review should determine that:

1. Non-maintenance project work does not interfere with the basic maintenance program.
2. Manpower limits available for non-maintenance work are carefully assessed by maintenance before such work is accepted. If interference with the basic maintenance program is possible, the work is deferred.
3. An effective priority system exists for allocating labor against major maintenance versus non-maintenance projects.
4. Project engineers are available to plan and control major non-maintenance projects effectively.
5. The work order system controls non-maintenance projects effectively.
6. Information is available on the cost and performance of major non-maintenance jobs.
7. Major modifications are prohibited unless they are reviewed and approved by engineering.
8. Major non-maintenance projects follow specific funding and approval procedures.
9. Approved non-maintenance projects must have adequate instructions, drawings, diagrams, and so forth, before they are carried out by maintenance.
10. New installations done by a contractor are monitored to ensure their maintainability before they are put into service.

11. Maintenance crews are not left with inadequate engineering instructions or poorly coordinated on-site support for project work.

You must be involved in this potentially troublesome "free for all" that often threatens to undermine the harmonious interaction between operations, maintenance, and plant engineering.

MOBILE EQUIPMENT MAINTENANCE

Objective

An efficient mobile equipment maintenance program is essential to an operation in which mobile equipment represents an important segment of production or support work. Your review should help determine whether:

1. There is a well-organized area for units awaiting service.
2. Units are washed to ensure that maintenance is carried out on properly cleaned vehicles.
3. Preventive maintenance services are carried out in specifically designated areas.
4. No major repairs are attempted during preventive maintenance inspections.
5. Deficiencies found during PM inspections are reviewed as the basis for further repair actions.
6. Areas designated for PM services are not used for long-term repair activities.
7. The facility is divided up into areas for PM, running repairs, and long-term repairs.
8. The facility is organized to handle the maintenance of different types of units effectively.
9. Major maintenance is properly planned and scheduled.
10. PM services (inspection, lubrication, and other servicing) are scheduled far enough in advance to allow operations to easily comply with the schedule.
11. Operations complies fully with the PM schedule and its compliance is measured.
12. Warehouse facilities are conveniently located and efficiently run.
13. Lube lines, compressed air lines, welding gas lines, and so forth, are conveniently laid out.

14. Areas for welding, tire repair, component rebuilding, and so forth, are properly located.
15. Garage configuration has taken into account the need for design features such as drive-through bays, bay doors of sufficient height, shop ventilation, heating for winter, cold-weather plug-ins, drainage, and so forth.
16. Internal communications such as integrated telephone-loudspeakers operate effectively to ensure good communication despite loud, distracting shop noises.
17. Completed work is picked up quickly by users from designated "ready areas."
18. There is evidence of an effective quality control effort like road testing.
19. Field repair of mobile equipment is carried out effectively.

With some exceptions, mobile equipment maintenance is separated from plant activities and it tends to get less management attention. Make a special effort to look in periodically using the points suggested previously as a guide. It will pay dividends in keeping mobile equipment maintenance personnel effective.

Summary

The most frequent cause of poor execution by maintenance departments is the failure to define their program effectively and educate personnel. Not surprisingly, maintenance personnel who have little idea of what is expected of them can hardly perform adequately. Similarly, operations is at a loss of what to expect. Therefore, a personal look at the quality and completeness with which your maintenance department has defined its program can assure this essential task is done well. As a result, better performance from maintenance can then be expected.

15

Performance: Assessed and Improved

What key elements of maintenance performance should you assess to ensure that they are continuously improved and will ensure maintenance effectiveness?

WORK CONTROL

Definition

Work control ensures that quality work is being done on time. Work control procedures begin with proper direction from the superintendent. They continue with effective planning and scheduling of major jobs. The most critical application of work control is exercised by the first-line supervisor. This supervisor assigns work, supervises crew members, coordinates outside supporting resources, ensures timely job completion, and is ultimately responsible for the quality of the work performed. Work control procedures should make it possible for craft personnel to work effectively with minimum supervision and serve as the basis for guiding successful team response to maintenance needs. Your review should determine that:

1. You have provided a policy that specifies effective work control.
2. First-line supervisors assign work effectively.
3. Major jobs are planned with efficient activity sequences.
4. Major jobs start on time, remain on schedule, and are completed on time.

5. Standards are applied to selected major jobs.
6. Job cost and performance are used to assess job progress.
7. Variances such as excessive labor or job cost must be explained.
8. Supervisors have adequate time to control work and craft personnel have adequate time to do work.
9. Supervisors can carry out and control several major jobs simultaneously using effective control and delegation techniques.
10. Several crafts working together (even when some are not regular members of a crew) are well controlled.
11. Mobile equipment use, rigging, transportation, and so forth, are carried out effectively.
12. As jobs are completed, maintenance advises production and invites inspection of work.
13. Repetitive work is examined at regular intervals for improvements in tools or equipment used and procedures followed.
14. Front-line supervisors evaluate worker's performance.
15. Verbal instructions can be used without loss of control or information.
16. Quality control is strongly adhered to.
17. There is seldom any loss of control of work.
18. First-line supervisors control work effectively.

Work control requires direct field observations up to the point of a manager assuring himself/herself that supervisors are doing it competently. Often, otherwise competent supervisors do a poor job because of too many meetings or administrative tasks. Other supervisors may do poorly because they are simply not "people-oriented" and end up hampering a crew that works well together. Regardless of the reason, it is incumbent on the manager to look, learn, and act in the most vital area of maintenance performance—getting the work done.

LABOR CONTROL

A Fact of Life

The only direct way that maintenance can control cost is the efficiency with which it installs materials. This makes control of labor a vital function. Therefore, management actions to improve the quality of labor control contribute to maintenance cost control. Thus, your assessment of labor control should determine whether:

1. You have provided a policy that emphasizes the control of labor.
2. Absenteeism is well controlled and there is a fair but well enforced vacation policy.
3. Overtime is properly authorized and controlled.
4. Actual use of labor by type of work performed is reported and analyzed to check the effectiveness of labor control.
5. Planning and scheduling procedures emphasize accurate labor estimates and allocation of available manpower.
6. The amount of labor to be used on important planned jobs is specified and actual use is verified.
7. Personnel can be shifted between production areas to meet peak work loads.
8. On the completion of major jobs, actual labor use is analyzed to help assess the quality of labor control.
9. Individual first-line supervisors are held accountable for the effective use of their crews.
10. First-line supervisors assign work and control labor effectively and poor use of labor must be explained by supervisors.

Regrettably, effective control of labor is seldom an outstanding trait among maintenance supervisors. In fact, many are zealous in the opposite direction—if a crew of 9 will do the job, 14 are better. Obviously, with more people than are needed, there will be less incentive to control work and improve productivity. Thus, your labor costs will be excessive. The only ways you can reduce the cost of maintenance is to do it with fewer people and do it less frequently. Thus, achieving effective labor control often means reducing work force level. Do not count on senior maintenance personnel to do this for you; they find it an uncomfortable experience. Checking the quality of labor control is an activity that you must be directly involved in—if you want results.

PRODUCTIVITY

Definition

Productivity is the measure of the effectiveness of labor control. It is the percentage of time that a maintenance crew or individual craft personnel are at the work-site, with tools, performing productive work. Maintenance is no more likely to measure productivity than they are to unilaterally reduce the work force size. They need to be encouraged, strongly, to do so. To attain productivity improvement you should establish that:

1. You have provided a policy prescribing that productivity checks will be carried out regularly and improvement targets established.
2. The hourly work force understands productivity measurements.
3. Maintenance management and its supervision has carefully explained why productivity checks are necessary to reduce hourly work force fear of layoffs or work force reductions.
4. Productivity is measured using logical indices such as percentage of hours working, traveling, idle, waiting or performing clerical tasks.
5. Useful, nonthreatening techniques are used to measure productivity that will get the job done without resistance.

If you want better labor control you must measure productivity. However, your personal involvement will be required if it is to be accomplished initially. If you can use techniques that successfully improve productivity without maintenance resistance, maintenance may then respond with more interest but seldom with real enthusiasm. This is an issue that without your involvement it simply will not get done.

COST CONTROL

An Explanation

Maintenance costs are forecasted against production targets relative to time periods: monthly, annually, and so forth. This budgeting process establishes expectations against which actual expenditures are compared. With such a yardstick, performance is measured by comparing man-hours used, labor cost accrued, material cost accumulated, and total costs committed. The focus of these costs is equipment, buildings, or facilities maintained, activities performed, or jobs completed. Maintenance controls costs primarily by the effective use of labor. They can influence costs indirectly by encouraging production to operate equipment correctly. The overall cost control effort should include adequate preventive maintenance, emphasis on planned work, and the use of cost-related information to help anticipate the need to prepare for and plan major work. Your review of the adequacy of cost control measures should help determine whether:

1. A logical budgeting technique is used.
2. The budget is based on production targets and their relationship to maintenance costs.
3. Capital expenditures and overhauls are budgeted and overhauls subjected to cost and performance review.
4. There is a criteria for justifying overhauls versus equipment replacement.
5. A cost summary compares actual and budgeted costs by cost-center on a month and year-to-date basis.
6. Maintenance costs can be isolated for equipment and components.
7. Supervisors are required to explain excessive cost variances.
8. Supervisors have cost-related performance targets such as control of overtime, tool-loss reduction, accident prevention, or productivity improvements.
9. Operations cooperates fully in ensuring that maintenance services can be carried out at least cost.
10. There are regular, formal cost review meetings held by maintenance and operations.
11. Maintenance controls labor and improves productivity.
12. Maintenance reviews warehouse stock to check consumption rates and identify parts to be removed from stock or alter on hand balances.
13. Maintenance anticipates purchasing needs to avoid extra costs like air freight.

Successful cost control requires a work order–information system that provides information to judge performance and guide corrective actions. Your insistence that maintenance use the information to control its costs will be pivotal. You must then push operations to operate equipment correctly, report problems promptly, and ensure its equipment modification needs are realistic. In addition, policies on capitalization, overhaul, and equipment replacement must underscore these efforts.

HOUSEKEEPING

Objective

Good housekeeping practices have the objective of keeping the plant environment in an orderly, attractive condition while helping to preserve assets. They result in the creation of a working environment in which personnel are motivated to work more effectively. An effective barometer of the quality of maintenance is the degree to which they pick up after themselves. Thus, your assessment of maintenance should assess whether:

1. Monthly housekeeping inspections are conducted.
2. Maintenance tools, materials, and debris are removed promptly from job-sites once work has been completed.
3. Housekeeping responsibilities are accepted by production personnel and carried out as a regular part of their jobs.
4. There is concern for housekeeping as evidenced by a neat, clean, and tidy appearance everywhere.

An occasional trip through the plant and direct conversation with personnel to let them know you care are often sufficient to improve housekeeping practices.

SAFETY

Objective

Safety is a primary consideration in the conduct of maintenance. Quality safety training and adherence to proper safety practices help to create an environment in which maintenance can carry out its work with confidence that accidents and injuries can be prevented. Your link with safety should extend to field observations as well as administrative indices of safety performance. Thus, you may be able to better anticipate potential hazards and avoid them. Your safety assessment should determine whether:

1. Every maintenance employee attends at least one safety meeting per month.
2. Maintenance has continuously improved its safety record.

3. All maintenance personnel use prescribed safety equipment such as safety glasses, boots, and hard hats.
4. Maintenance workers never work on a job without proper safety clearances.
5. All maintenance personnel are well trained in the safety considerations related to their jobs.
6. Fire protection is effective.
7. Vessel entry procedures are carefully followed.
8. Welding and burning permits are followed.
9. There is involvement and commitment to safety.

Safe work practices are more a frame of mind than a set of rules. Thus, the worker who wishes to avoid injury will agree with every rule. But, more to the point, the worker will follow them. Therefore, your physical presence and the forceful identification of unsafe practices will not only reinforce your commitment in the eyes of workers, but they will recognize your priorities of safety over production targets.

Summary

Maintenance performance at the management level is often appraised through the use of performance indices such as cost per unit of product, labor cost to install each dollar of material, and adherence to the budget. While these are helpful, it is better to assess the basic field performance first and determine whether the vital issues of work control, labor control, productivity, cost control, housekeeping, and safety are being properly looked after.The manager's review of selected maintenance performance areas often reveals the logic and commitment behind the maintenance effort to control and improve performance. As a result, your ability to guide corrective actions is enhanced with personal knowledge gained in selectively examining the most critical issues suggested in this chapter.

III

What Evaluation and Improvement Strategy Will Ensure Successful Maintenance Performance?

16

Evaluation
and Improvement Strategy

How can you develop a strategy for evaluating maintenance and convert
improvement needs into a successful plan of action?

EVALUATIONS IMPROVE
MAINTENANCE PERFORMANCE

Performance = Profitability

Quality performance by maintenance helps assure plant profitability.
Conversely, when maintenance performance is poor, costs will be excessive and unnecessary downtime will cripple plant efforts to meet
production targets and stay within operating budgets. Unfortunately,
desired performance improvements are often not realized because the
evaluation of maintenance, the first step of improvement, is not done
adequately. Too often, it is not done at all. Perhaps you have observed
at your plant that production gets a lot of attention. Their performance
is evaluated frequently, in detail, and usually very effectively. You may
wonder why the same attention is not given to maintenance. Often it is
because plant managers, who insist on evaluating production operations, may fail to realize that maintenance performance requires the
same emphasis. Without evaluations the specific improvements in
maintenance performance that could help assure profitability are often
never identified. Support for maintenance from production, purchasing, and so forth, has significant influence on maintenance perfor-

mance. Unfortunately, maintenance has little control over how much support they provide or its quality. Thus, any evaluation of maintenance will be more acceptable to them if you also evaluate those supporting aspects of plant operation and administration that affect maintenance performance.

ACHIEVING POSITIVE ATTITUDES TOWARD EVALUATIONS

Attitudes

A primary reason for poor attitudes toward evaluations is misunderstanding. This leads to suspicion and causes resistance. For example, maintenance may resist an evaluation claiming it is disruptive to workers. Yet, the evaluation may reveal that poor work control techniques are already causing excessive disruption of work. Also, warnings of unpleasantness with the union may be hinted by maintenance to create uncertainty about the timing of an evaluation. However, the evaluation may reveal circumstances that the union desperately wishes be identified and corrected. Under such conditions, you may not go ahead with the evaluation. Thus, problems that should be corrected are never identified.

Resistance

Maintenance resists evaluations because there are too many factors over which they have no control that influences their performance. For example, the warehouse may be poorly run because it is managed by a remote, disinterested accounting department. Equipment may be unreliable because it is being pushed beyond maintenance due dates by operators pursuing unrealistic production targets. Resistance must be eliminated by removing any threat to maintenance implied by an evaluation. Therefore, the evaluation must distinguish between those factors over which maintenance has responsibility and those over which it has no control like the unresponsive warehouse run by an indifferent accounting department.

Responsibilities for performance over each group of factors must be clearly established. Thus, if poor maintenance performance is due to poor warehouse support, this must be brought out into the open and corrected. Maintenance can be evaluated fairly only if such aspects as operations cooperation, staff department support, and management sanction are evaluated as well. Maintenance, after all, is a service. They cannot, for example, force operations to comply with their program nor can they demand high service levels from the warehouse. Yet, they need this cooperation and support to succeed. However, once it is clear that the factors over which they have little control will also be evaluated, most resistance to evaluations will disappear.

Maintenance also resists evaluations because of the limited management background of many of its key people. Its leaders rarely come from outside of maintenance. Rather, they are usually inbred and carry their management styles from the supervisor to superintendent level unchanged. Many are graduates of the "school of hard knocks." University degrees, while not a requirement, are rare among maintenance leaders when compared with operations, for example. Few have had any management training. They assume their supervisory responsibilities with little preparation for the challenges that they will face. Some may even be unwilling supervisors and uncomfortable in the role. The majority are ex-workers personnel. While they are competent in a single skill, they are often unfamiliar with the management techniques necessary to meet their obligations as supervisors. Further compounding the situation is the fact that a large number, as ex-workers, are former union employees who may harbor an "us versus management" viewpoint. Thus, the education of maintenance supervision on the necessity of evaluations—as the first step of improvement—will be essential to win their support.

DISPLACE RESISTANCE WITH EDUCATION

Policy

The education process should be initiated by establishing a policy that maintenance be evaluated regularly. This step redirects energy that could be spent in resisting into preparing instead. Once you do this you can then establish the why, what, when, and how of evaluations with a much more receptive audience.

Purpose

The purpose of the evaluation must be carefully explained before the requirement for an evaluation can be imposed. It should be explained that the evaluation is a checklist describing what maintenance should be doing. Its results should describe how they did and provide a basis for developing an improvement plan. It must be promoted as an effort to identify what is done well, not what is done poorly. It is your opportunity to help maintenance account for its support of plant profitability objectives.

Evaluation Content

The content of the evaluation must also be explained. A complete, valid evaluation must include a look at the maintenance organization, its program, and, inevitably, the environment in which maintenance operates. The main points are:

1. The Organization
   a. Labor control and productivity
   b. Supervision
   c. Work load measurement
   d. Motivation
   e. Craft training
   f. Supervisor training
2. The Program
   a. Program definition
   b. Terminology
   c. Preventive maintenance
   d. Work order system
   e. Information
   f. Planning and use of standards
   g. Scheduling
   h. Work control procedures
   i. Maintenance engineering
   j. Use of technology
3. The Environment
   a. A clear objective
   b. Adequate policies
   c. Management support

d. Production cooperation
e. Staff department service
f. Effectiveness of material control
g. Observance of safety procedures
h. Housekeeping

The environment is the attitude of the plant toward maintenance. When poor, it is the most common reason for resistance to evaluations. Even a first rate maintenance organization with an effective program can fail if the environment does not provide receptivity and support. Examining the environment includes answers to some hard questions about the quality of management sponsorship, the degree of operations cooperation, and the level of service provided by staff departments such as purchasing or stores. Not surprisingly, when maintenance realizes that key factors over which they have little control—but which influence their performance—are being evaluated fairly, their resistance diminishes. Be aware that an evaluation of the plant environment can reveal an uncooperative operations department or an unsupportive purchasing agent. Therefore, it will be necessary for you to condition these other departments for their evaluation while also requiring their accountability for support of plant profitability objectives.

Timing

Timing of evaluations is important. If they are conducted on a regular, continuing basis, people will look forward to them as an opportunity to demonstrate progress. If maintenance feels that the previous evaluation was constructive, they will prepare for the next one without hesitation. They will also make a conscientious effort to improve on previous results.

MAINTENANCE EVALUATION STRATEGY

A Ten Step Maintenance Evaluation Strategy

To take advantage of the improvement steps that evaluations offer, an overall strategy is necessary. The strategy must identify actions to organize and conduct an evaluation and convert the results into improvements. A ten step strategy includes the following points:

1. Develop a policy on evaluations
2. Provide advance notification
3. Educate personnel
4. Schedule the evaluation
5. Publicize its content
6. Use the most appropriate evaluation technique
7. Announce results
8. Take immediate action on results
9. Announce specific gains
10. Specify the dates of the next evaluation

This strategy recognizes that maintenance has the potential of contributing significantly to profitability. Maintenance will not evaluate itself. You must acknowledge that to some maintenance departments an evaluation is like a self-inflicted wound. You must be the catalyst to ensure that an evaluation takes place. Once the evaluation has taken place, you must swiftly convert the results into a plan of action and exert the leadership to develop and implement needed improvements. If you have done a good job, the evaluations will correctly identify needed improvements and provide you with the opportunity to initiate corrective actions. The evaluation is the first of the steps to achieve plant profitability through improved maintenance performance.

Step 1: Develop a Policy on Evaluations.   There must be a strong policy requiring that maintenance be evaluated on a regular, continuing basis. Such a policy will preclude the resistance that many poor maintenance departments exhibit to discourage evaluations. This policy will help to redirect the energy of resistance into an effort to prepare for evaluations instead. In a typical evaluation policy, the maintenance organization, its program, and performance will be evaluated on a regular, continuing basis to identify areas of improvement and act on them. Evaluations will also identify effective areas and advise personnel of their accomplishments. Because successful maintenance depends on operations for cooperation and staff departments for support, these aspects will also be evaluated.

Step 2: Provide Advance Notification.   Advise personnel about the evaluation and make a preliminary statement about its content, purpose, and utilization of the results. Eliminate surprises with the announcement and emphasize the policy of regular, continuing evaluations. The announcement should consist of the following six points:

1. Provide the dates of the evaluation.
2. State the objectives.
3. Remind personnel of the results of previous evaluations, commenting on accomplishments and further improvements needed.
4. Explain how this evaluation is in line with the long-range policy of ensuring continuous improvement.
5. Thank everyone in advance and state that you are looking forward to their help.
6. Offer to meet with anyone who may have problems with the evaluation content or dates.

Step 3: Educate Personnel. Emphasize the positive aspects of the evaluation in the educational effort. Change unfavorable misconceptions of evaluations by telling personnel that they are the means of finding out how maintenance can do better. In preparing personnel for evaluations, acknowledge that there may be a genuine fear of audits and evaluations. Some fear is based on previous bad experiences while some fear has no basis. Many simply do not want anyone "looking over their shoulders."

> One maintenance manager, thinking it might be funny, proclaimed that "Auditors were those who come in after the battle is lost and bayonet the survivors!"

Such attitudes contribute little to the mental preparation that must be made if a plant is struggling to improve. Correct such unfavorable attitudes with positive thoughts. An evaluation is the first step of improvement. Some unions have been known to threaten strikes when an evaluation is announced. They fear that their "sacred promise" of preserving jobs will be threatened by possible work force reductions as the result of an evaluation. The fact that the evaluation can also lead to improvements that will make their work easier seldom occurs to them. It follows that maintenance supervisors who are caught up in the threat of their jobs being made harder by the union position may resist as well. Thus, direct the educational effort toward both supervisors and workers.

No Surprises

Avoid surprising personnel as this adds to resistance and creates distrust. Adults do not like surprises. Therefore, let personnel know what is coming and avoid resistance. The evaluation should be considered as a checklist describing what maintenance should be doing. Its results describe how they did and provide a basis for developing an improvement plan.

Evaluating Dual Operations

Resistance to being evaluated occurs also when there are several operations that will be evaluated. The concern is that performance between several maintenance operations will be unfairly compared when, in fact, they may not be similar and would be better compared on another basis. Regardless, comparisons will be made. Therefore, be aware that the evaluation sponsor, a general manager, for example, must convey a supportive attitude. If you are that general manager, provide encouragement to conduct the evaluation in the first place and then follow up to see that something constructive is done with the results. Offer to help achieve the improvement needs identified with tangible resources like funds for an additional mobile crane. Get involved in the education process by making your position clear well in advance.

In your own plant, you will want to know how well your policies are understood and how effectively the procedures based on those policies are being carried out. While the first concern is the quality of maintenance performance, you should be equally interested in learning how well, for example, production supports and cooperates with the maintenance program. During the educational process, reinforce your policy about evaluating those areas outside of maintenance, like material control, that impact their performance.

Step 4: Schedule the Evaluation. Schedule the evaluation carefully, particularly if it will involve a physical audit lasting several weeks. In selecting the evaluation dates, be aware of potential conflicts which might distort evaluation results. If a shutdown has just been completed there may be distracting start-up problems. Similarly, if a shutdown is coming up, preparation may compete for the attention of personnel. Peak vacation periods may find key personnel away from the plant. Be aware of the expected absences of key people and weigh their nonparticipation in deciding when the evaluation should be conducted. Recent personnel changes could limit knowledge of evaluation points. Similarly, personnel cutbacks or staff reductions might affect attitudes. In general, the evaluation should be carried out in a stabilized situation with as few distracting conditions as possible. With suitable advance notice, the plant can prepare for the evaluation and look forward to learning how it is doing.

Step 5: Publicize Its Content. The content of the evaluation should be defined in advance. A maintenance evaluation covers a variety of activities. Therefore, it is unlikely that any advance notice will constitute a dramatic shift in performance. By announcing what is to be evaluated you will instead help organize the evaluation. Reports will be ready, personnel can be scheduled for interviews, and the evaluation can be carried out more effectively. Agreement on evaluation content also clarifies expectations of both the maintenance department and the evaluators. For example, if productivity is to be measured, hourly personnel should be advised so that they do not misunderstand the intent.

The scope of the evaluation should include three broad areas.

1. The organization
   a. The efficiency of the maintenance organization in controlling its personnel and carrying out its work.
   b. The quality of maintenance supervision and its effectiveness.
   c. The ability of maintenance to measure its work load and determine the correct work force size and craft composition.
   d. The degree to which maintenance effectively controls labor utilization.
   e. The degree to which maintenance considers productivity important and acts to improve it.
   f. The level of motivation of maintenance supervisors and workers.
   g. The quality and effectiveness of craft training.
   h. The quality and effectiveness of supervisor training.
2. The program
   a. How effectively maintenance has defined its program so that everyone can act in a well-informed manner.
   b. How well maintenance terminology has been defined, communicated, and used correctly.

c. The effectiveness of preventive maintenance in extending equipment life and avoiding premature failures.

d. The quality and effectiveness of the work order system—the communication link of maintenance.

e. Whether maintenance has the necessary information to identify work, control its performance, and measure its effectiveness in reducing costs and minimizing downtime.

f. The quality of maintenance planning in assuring major jobs are carried out effectively with minimum resources and least elapsed downtime.

g. How well performance standards are used to ensure job quality and control labor use.

h. Whether scheduling assures the least interruption of production and the best use of maintenance labor.

i. The degree to which work control procedures assure effective labor, quality work, timely work completion, and knowledge of job status.

j. How effectively maintenance engineering techniques are used to ensure equipment reliability and maintainability.

k. The degree to which modern technology is employed by maintenance to help perform its tasks more effectively.

l. How well maintenance uses backlog data to determine whether they are keeping up with the generation of new work and if the current work force has the capacity to meet the work load.

3. The environment

a. Whether management has assigned a clear, realistic objective to maintenance.

b. The degree to which maintenance is supported by policies that ensure common understanding and compliance with most procedures.

c. How well management supports the maintenance program and causes it to be used effectively in the support of the production strategy.

d. How well production understands the maintenance program, cooperates, and uses its services effectively.

e. Whether staff departments such as purchasing understand the maintenance function and support it with a high level of service.

f. The effectiveness of material control in support of the maintenance function.

g. The degree to which safety procedures are followed and unnecessary accidents or injuries avoided.

h. How well maintenance carries out housekeeping tasks to help preserve assets and create a neat, clean working environment.

Examination of the environment includes steps that you must take to ensure the success of the maintenance program. Typically, they include a clear objective and realistic policies to guide other departments in how they must support maintenance.

Step 6: Use the Most Appropriate Evaluation Technique. Evaluation techniques used depend on the plant situation and the evaluation objectives. While some operations may require an evaluation in which every detail must be scrutinized, other operations, having established the essential pattern of evaluations, may simply check progress by measuring only a few critical areas. There are three techniques that may be used for the evaluation of maintenance:

1. Physical audit
2. Questionnaire
3. Combined Audit and Questionnaire

Physical Audit

A physical audit is usually conducted by a team. Generally, the team is made up of consultants or company personnel or both. They examine the maintenance organization and its program as well as activities which affect maintenance such as the quality of material support. The physical audit examines the total maintenance activity first hand. It includes observation of work, examination of key activities such as preventive maintenance, planning, or scheduling, review of costs, and even the measurement of productivity. It relies on interviews, direct observation of activities and examination of procedures, records, and costs. When done properly, the physical audit produces effective, objective, and reliable information on the status of maintenance.

The physical audit by itself should be used when you feel that personnel could not be frank, objective, and constructive in completing a questionnaire. If the physical audit technique is used, be prepared to spend several weeks undergoing the evaluation. The evaluation can be disruptive because of the time required by personnel to explain procedures or participate in interviews. Therefore, it must be well organized in advance.

Questionnaire

When using a questionnaire a cross section of randomly selected plant personnel including management, plant staff departments such as purchasing, production, and maintenance are given an opportunity to compare their plant's maintenance performance against specific standards. Appendix F, Typical Maintenance Questionnaire, illustrates the type of standards against which preventive maintenance, for example, should be evaluated. While a questionnaire is subjective, the results are, nevertheless, an expression of the views of the plant personnel. Therefore, participants have committed themselves to the identification of improvements they see as necessary. To most of them, this constitutes potential support for the improvement effort that must follow.

A cross section of about 15% of the plant population is adequate to produce good results. Typically, the groups would include maintenance, operations, and staff personnel (like accounting). Within each group there should be a vertical slice of the organization. For example, the maintenance group might include the superintendent, several supervisors, the planner, and a few craft personnel. Thus, a procedure developed by the superintendent would be assessed by those trying to make it work. Participants should have personal knowledge of maintenance performance for the standards against which they are comparing maintenance. Further, there should be selectivity in who responds to what. The accounting manager, for example, could evaluate the quality of labor data reporting, but he could not evaluate housekeeping. Use care in administering the questionnaire to ensure that personnel are qualified to respond. Those participants outside of maintenance, for example, must respond only to those standards on which they have personal knowledge. An equipment operator might be unhappy with the quality of servicing, but he may be totally ignorant of the content of the service checklist and service frequency. Thus, he can state his displeasure only on the quality of service he received, but not on the quality of the overall program. The evaluation must be administered in an environment in which questions can be answered to clarify evaluation points. Those personnel administering the questionnaire should brief participants in advance and be available during the completion of it to answer questions or interpret standards. When carried out properly, the questionnaire can produce reliable results quickly while minimizing disruption to the operation.

The questionnaire has the advantage of being administered often so that progress against a "benchmark" can be measured. For example, one plant was able to quickly establish areas in a second evaluation in which improvement was still needed while setting aside areas in which good progress had been attained since the initial evaluation. The questionnaire is the best choice when a quick, nondisruptive evaluation can serve as a reasonable guide in developing a plan of improvement action. It must be carefully crafted so that it embraces all of the elements of the maintenance program like PM as well as those activities which impact the program like purchasing.

Questionnaires are rarely of value if they are not administered in a controlled environment. If distributed for completion at the respondents' leisure, expect poor results because there are too many opportunities for misunderstanding. Administer questionnaires in a controlled environment where participants can be oriented and their questions answered to ensure understanding of the points that they are asked to evaluate.

Physical Audit and Questionnaire

The combination of a physical audit and a questionnaire provides the most complete coverage. The techniques work together; the questionnaire provides confirmation of physical audit findings, for example. This technique is preferred by consultants and corporate teams because it combines the objectivity of the outsider in the physical audit while the questionnaire helps to educate personnel and gain their potential early commitment to improvement. As the insiders, their help is needed. Without it, little will happen. This combined technique is the best way of preparing for the improvement effort that must follow.

Step 7: Announce Results. By publicizing the evaluation results, there will be clear evidence that you acknowledge both the good and the bad. More importantly, you also exhibit a commitment to do something. By sharing the results of the evaluation, you confirm that you expect the help of your personnel in attaining improvements. Never keep the results a secret as this will decrease credibility and make improvement actions more difficult. Evaluation results should be discussed openly and constructively so that the personnel who must later support improvements are being brought along. Do not allow the maintenance superintendent to rationalize that he is solely responsible when the evaluation reveals a poor performance. He probably got a lot of help from many people in the resulting poor performance. He cannot, for example, influence the blame placed on maintenance for unreliable vehicles when, in fact, operations rarely made the mobile equipment available for scheduled PM services or excessive requests from operations for unnecessary equipment modification that misuses maintenance manpower.

Little is gained by trying to fix blame. Maintenance reaches into so many areas that few people would be without some degree of responsibility for the decline of maintenance performance. Always look ahead—determine the current performance level and move forward from there. Maintenance, after all, is a service. They cannot compel operations to follow the program. Only you, as manager, can direct cooperation from operations and support from staff departments. Work toward creating an environment which assures cooperation and support. By including operations and staff departments in the evaluation, you not only examine aspects over which maintenance has little control, but you identify what these departments can do to help. Share the results openly. Plants performing poorly often do not show results to the personnel who participated. Better plants not only share the results but seek help in interpreting the results and soliciting recommendations.

> One successful maintenance manager observed that "whatever the current performance is, it didn't get that way overnight. We must all work toward improving it."

He approached his task of improving maintenance performance by saying,

> "Since this is what *we* think of *our* maintenance program then let *us* now consider what *we* must do about it."

That plant was already on its way to improving because the support and enthusiasm for doing better had been successfully harnessed even before the evaluation was completed. There was involvement and it showed.

Deal with the Results Fairly

As plant manager, the evaluation could well reveal that you have not:

1. Established a clear objective for maintenance.
2. Provided policy guidelines on which they can build day-to-day procedures.
3. Required a definition of the maintenance program.
4. Defined your production strategy.

Similarly, production personnel may be guilty of calling every "squeak" an emergency. Also, the purchasing agent may not be as service oriented as he might. If such circumstances come to light, own up to and correct them. Not surprisingly, evaluations can reveal facts about maintenance that were concealed by misleading information or defensive attitudes of maintenance personnel. Thus, if the maintenance superintendent acknowledges a program not adequately defined or a need to align basic program elements like PM, planning, scheduling, and maintenance engineering, give him/her unqualified support.

Step 8: Take Immediate Action on the Results. The most convincing way to demonstrate that the evaluation was a constructive step is to organize an improvement effort immediately. You must commit to a constructive use of the results by converting them into an improvement plan and immediately organizing the improvement effort. This is the main objective of the evaluation. If the evaluation is one of a series, results should be compared with the previous evaluation. This demonstrates progress as well as the identification of areas that need more work. Separate the good from the bad. Offer congratulations on the good performances and organize the activities requiring improvement into priorities. Actively solicit help from anyone capable of providing it. Most will participate willingly. Thereafter, a plan for further improvement should be developed and corrective actions implemented. If there are corrective actions beyond the capability of maintenance offer or get help. Senior managers are often delighted to be asked to help and are greatly relieved to find an evaluation initiative they did not have to force on a plant. Set up an advisory group and get underway. Let them determine first why certain ratings were poor. Then, ask for recommendations for improvement. Change the members of the advisory group frequently to encourage different views. As recommendations are made, try them in test areas before attempting plant-wide implementation. Appendix G, Action Group and Decision Group, suggests a method for converting evaluation results into realistic corrective action.

Step 9: Announce Specific Gains.   As soon as any gains that can be attributed to the evaluation are identified, they should be announced and credit given to the appropriate personnel. People like to know how they did. Tell them. In the process, your candor will invariably encourage a greater effort in future evaluations.

Step 10: Specify the Dates of the Next Evaluation.   As necessary, identify any additional activities that will be evaluated. Establish new, higher performance targets for the next evaluation. Reinforce your policy of continuing evaluations.

USING EVALUATION RESULTS
TO IMPROVE MAINTENANCE

Case Studies

There are useful lessons in observing how maintenance organizations have successfully converted evaluation results into improved performance. Case studies describe evaluation results and their conversion into corrective actions.

Supervisor Training

One of the areas that evaluations most frequently find in need of improvement is supervisor training. Most plant personnel described a nonexistent supervisory development program. There were major problems. Newly appointed supervisors were expected to figure out their duties for themselves. Training on the maintenance program was often vague because the program itself was seldom well defined and documented. Supervisors identified for promotion were seldom trained on the broader duties that they would be required to perform. There was little time allocated for supervisory training of any kind. In the little training given, emphasis was given to technical skills, rather than "foremanship." As a result, many supervisors were reluctant to move from their "one craft" crews to the "multi-craft crews" necessary for more flexible organizational arrangements like area maintenance. There was an impression that management saw supervisors as those who "caused equipment to be fixed" rather than those who managed the efforts of craftsmen in carrying out the maintenance program. Generally, supervisors were criticized for their lack of supervisory skill. Few had ever received any supervisory training. They were rated well on the technical competence which they carried into their supervisory jobs as they moved from craftsmen to supervisors. Most supervisors displayed a tendency toward working directly with their crews rather than leading them. Generally, superintendents were criticized for acting more like foremen than managers. In turn, supervisors were criticized for acting like workers. Supervisors were often labelled as "tire kickers" rather than managers. Poor marks were given for control of work and utilization of labor.

Solution:
1. Establish a policy for supervisor training which emphasizes foremanship.
2. Verify that there are no impediments precluding attendance at training such as an inadequate number of supervisors.
3. Require completion of the prescribed curriculum and then assess supervisory performance.
4. Give special emphasis to the training of new supervisors.
5. If in-house training is not available, arrange attendance at suitable commercial training.
6. Get involved in setting up the training, monitoring it, and checking progress.
7. Do not accept excuses for nonparticipation in the program.
8. Reconsider supervisor selection criteria and look for talent among those without inbred maintenance backgrounds.
9. Require emphasis on foremanship in supervisor training.
10. Demand that supervisors utilize crew members more effectively on repairs while they assume more responsibility for work control.
11. Initiate a supervisory selection criteria so that those with limited management talent will not stagnate the line of progression with people who are "dead-ended" when they reach the supervisor level.

Aligning Craft Training

By contrast with supervisor training, craft training programs were rated well and considered successful. Contractual requirements that training for hourly craft personnel be provided is often credited with their greater success. However, some craft training tended to emphasize what craftsmen felt they needed rather than what the maintenance situation required. One plant learned that crafts that would never use infrared testing techniques were receiving training on it instead of what their program required. Few craft training programs provided training on the overall maintenance program as opposed to skill training alone. As a result, many maintenance departments attempting to convert to a team organization found the transition difficult.

Solution:
1. Establish craft training needs before the program is organized.
2. Require regular reports on craft training progress and periodically review its content to ensure that it remains consistent with needs.
3. Attend a sampling of the training to demonstrate personal interest.
4. Talk with craftsmen to determine their level of satisfaction with the training.
5. After defining the overall maintenance program ensure that it is made a regular part of the curriculum of all training, and not craft training alone.
6. Periodically, have a training needs analysis conducted to ensure continuing adequacy of the program.

Define the Maintenance Program

Few maintenance departments had solid, well-documented, and publicized definitions of their programs. As a result, neither the maintenance personnel performing work nor their customers understood the program. Program definition should explain how maintenance services are requested, organized, executed, controlled, measured, and so forth. There was confusion on how to do these basic things. Often, operating personnel acknowledged no role other than to submit work requests. There was little joint scheduling of major jobs and confusion on what was to be done and when. Performance on major jobs was seldom questioned by operations nor were overall maintenance costs. Results usually indicated that since maintenance did not advise anyone how they operated, the misunderstanding was expected.

Solution:
1. Verify that maintenance has defined its program.
2. Ensure that the program has been explained to maintenance personnel as well as to their customers.
3. If the program is not defined, require that it be defined, documented, published, and personnel educated.
4. Follow up to verify that the program definition is understood and the program is followed.

Develop the Maintenance Objective

Many maintenance departments had vague objectives like "support operations" or "get the product out." Those examined were compared with a reasonable standard such as:

> The primary objective of maintenance is to maintain equipment, as designed, in a safe, effective operating condition to ensure that production targets are met economically and on time. Maintenance will also support non-maintenance project work (like construction) as the maintenance work load permits. In addition, maintenance will maintain buildings and facilities and provide support services such as hoist-operation or power-generation.

Generally, the absence of a clear maintenance objective confused the maintenance program. Typically, maintenance, instead of focusing on preserving assets and ensuring dependable equipment, was too often diverted into non-maintenance activities such as performing "process optimization" (making the process better) or moving, upgrading, modifying, and installing equipment, and even performing construction work while basic maintenance was neglected.

There was confusion in differentiating between maintenance and non-maintenance work not only within maintenance but plant-wide. Maintenance is the repair and upkeep of existing equipment, buildings, and facilities to keep them in a safe, effective, as designed, condition so they can meet their intended purpose. Maintenance is an operating expense. Thus, non-maintenance like construction or equipment installation is generally capitalized, depending on the cost. Maintenance departments without clear boundary lines and specific policies, regarding the division of maintenance and non-maintenance work, usually failed to carry out the basic maintenance program adequately. Some maintenance managers with responsibility for maintenance helped create the problem by giving undue emphasis to non-maintenance project work.

Results further revealed that some plants had what appeared to be low cost maintenance programs. In reality, maintenance resources were misused on projects, often over the objection of the maintenance. Typically, deterioration of equipment resulted from a lack of maintenance and requiring its premature replacement. Often, an unfavorable maintenance budget was due not to excessive maintenance cost but rather to non-maintenance work that was expensed rather than capitalized.

Solution:

1. Review the current maintenance objective to ensure that it clearly establishes the conduct of the maintenance program as the departments primary responsibility.
2. If maintenance is required to perform non-maintenance work, ensure that there are safeguards to preclude improper utilization of maintenance resources.
3. Make sure the maintenance supervisors are not victimized by unrealistic emphasis of non-maintenance work like modification.
4. Verify that maintenance supervisors are not "construction types" who actively solicit non-maintenance activities (like construction or modification) while neglecting maintenance on existing equipment.

Establish Maintenance Policies

Maintenance, as a service, must support operations with a program to provide reliable equipment. However, operations must understand the maintenance program and utilize its services effectively. Uniformly, few managers clarified production responsibilities for utilizing maintenance services effectively. In a typical situation, an ambitious operating superintendent, in a quest for high production targets, presented a continuing need to management for modification of equipment. Having similar aspirations, management was sympathetic and approved most of his requests. However, it was soon discovered that many modifications were neither feasible nor necessary. Many should have been approved by engineering but were not. Some that should have been capitalized were chopped up into incremental expensed actions to avoid the "gauntlet" of getting capital funding approved. In the process, this operating superintendent, without malicious intent, diminished the maintenance program.

Ignoring maintenance schedules in the interest of meeting production targets also creates problems because maintenance cannot compel operations to follow the schedule. Such situations undermine the maintenance program. Evaluations determined whether there were policies which could prevent these situations. The results show that such policies, which can only be issued by plant management, were weak or missing. As a result, maintenance felt that operations viewed the maintenance program less seriously than they should have.

Solution:
1. Ensure that you provide policies to preclude abuses of the maintenance program.
2. Make operations responsible for the cost of maintenance. They will become more demanding of quality work which is completed on time and kept under budget. Expect a dramatic reduction in downtime.

Standardize Maintenance Terminology

Few maintenance departments adequately defined terms they used every day. Some terminology was misleading even within maintenance. The work load for maintenance could not be identified, much less measured. Especially confusing was the meaning of preventive maintenance, emergency repairs, and planned and scheduled maintenance. Few could distinguish overhauls versus rebuilds or modification versus corrective maintenance. In one plant, there were seven different versions of the meaning of PM just within maintenance. The absence of adequate definition of terms often emerged in other areas such as PM and planning where proper definitions were necessary to spell out effective control procedures or be able to measure the work load. Outside of maintenance there was confusion on basic terms.

Solution:

Check the existence of a basic terminology. If it does not exist, demand it. If it does exist, make sure it has been published and verify its correct use. Typical of the type of terminology that must be defined is the basic work that maintenance performs. See Appendix D.

Establish the Maintenance Work Load

Work load measurement was typically omitted or poorly done.

Work load measurement is the identification of the essential work to be performed by maintenance and the determination of a work force of the proper size and craft composition to carry it out.

Typically, one maintenance superintendent stated that he determined work force size by "adding workers until the overtime went down"—hardly adequate, but typical of the lack of regard for this matter. Results indicated that most maintenance departments did not know how to go about this task nor did they have elements in their information systems which permitted them to confirm or adjust current work force levels.

Solution:

Require maintenance to verify its work force size and craft composition regularly. If they are hesitant, chances are good that they have no idea how to do it. It is often necessary to demand that they do it, then guide them through some reasonable procedure. Using the definition of the work load as a starting point, estimate the amount of manpower for PM and planned and scheduled maintenance tasks. These can be computed. Then, make reasonable allowances for emergency repairs and unscheduled work. Next, identify the number of man-hours for routine activities such as training or shop clean-up. Assemble the results, comparing them with typical industry standards for manpower expenditure by type of work:

Preventive maintenance	10%
Planned and scheduled maintenance	50%
Unscheduled repairs	20%
Emergency repairs	10%
Routine activities	10%

Once these levels have been established with estimates, utilize the information system to confirm how labor is actually used and make corrections in the distribution. Then adjust the work force size and craft composition.

Measure Productivity

Evaluations revealed that worker productivity was seldom measured nor were steps taken to explain its purpose or importance. Hourly personnel feared that productivity measurements meant layoffs and work force reductions. Thus, they resisted, often to the point of threatening unpleasantness if management insisted. Most plants acknowledged that productivity was low and had not been measured, and that the current quality of labor control would not yield improvements. Invariably, the maintenance supervisor could have improved productivity by giving better work assignments, ensuring work was pre-organized, and spending more time supervising the crew. Why the supervisor did not was seldom investigated.

Solution:

Make productivity measurements mandatory. Techniques of random sampling are most effective when the pattern of what personnel are doing is identified. The percent of time spent working, remaining idle, travelling, waiting, and performing clerical functions needs to be identified. Also useful is when, during a shift, they are doing these things. If personnel, for example, are idle at the start of the shift it usually means that work assignment procedures are inadequate. Similarly, if early quits are the problem, it often means that the supervisor is overly involved in shift end administration. Left to their own devices, workers quit work early.

Productivity measurements are the most effective way of verifying the quality of labor control. However, it will be necessary to sit on most maintenance departments to make this happen. Yet, it should be done. A favorable first step can be taken by allowing maintenance to measure its own productivity—without incrimination. One maintenance department made major strides in improving productivity by measuring not productivity but instead the factors that inhibit work. The results indicated the degree to which factors such as the difficulty of obtaining materials robbed personnel of productive time. These results provided very specific improvement targets that responded to correction easily. Incidentally, the difference between measured delays and total time was the productivity. Appendix H, Measuring Maintenance Productivity, illustrates a technique in which maintenance personnel can measure productivity themselves. Find out what the supervisor is doing and how well he/she actually controls his/her crew. Supervision time and its quality is critical to improving productivity. He/she must spend at least 60% of his/her time on active supervision to assure 40% productivity (not very good; only 3.2 hours of work of each 8 hour shift). Therefore, make sure the supervisor has enough time to supervise. Educate him/her on the importance of improving productivity. Yet, if you have heaped administrative tasks on the supervisor, the poor productivity of the crew may be your fault.

Conduct Effective Preventive Maintenance

The biggest problem with Preventive Maintenance (PM) as revealed by evaluations is a failure to understand it. Essentially, preventive maintenance is inspection and testing to uncover problems in time to avoid emergency repairs and provide lead time to plan work; lubrication, cleaning, adjustment, and minor component replacement to extend equipment life. Views expressed during evaluations ranged from "everything you did before equipment failed" to "the whole maintenance program." This inconsistency carried over into the effectiveness with which it was able to be scheduled, executed, and controlled. Invariably, those who defined PM as routine, repetitive actions (inspection, lubrication, testing, and so forth) had the best control over its effectiveness. Those who did poorly in PM also did poorly in defining it. Many PM programs could not be administered effectively because inspection and testing were mixed with repairs. Deficiencies resulting from inspection and testing which should have been separated and classified into emergency or unscheduled repairs and planned maintenance were not. As a result, the identification of work load elements was blurred and precluded determination of the proper work force size and craft composition. Generally, good use was made of nondestructive testing (like vibration-analysis). However, such efforts were not always integrated into the overall PM program.

Solution:

Require that maintenance present their preventive maintenance program to operations and management. If they cannot, they do not have one. If they hesitate, they need to have it challenged or evaluated. Most likely, they may not know what to do or how to organize it. PM services like inspections for fixed equipment should be organized into routes in which equipment in each area is checked at fixed intervals. Inspections should be "detection oriented" and results converted into corrective actions. Completion of services should be verified. While this organizational concept is simple, it is often overlooked. PM is too important to ignore; your personal attention may be essential to make it successful.

Establish Effective Planning

Planning was generally not a well-executed function. Among the problems:

1. There were no criteria describing what work should be planned.
2. Planning procedures were poorly described.
3. Information for control of planned work was sparse.
4. There were not enough planners.
5. Planners were often misused on unscheduled or emergency work or as relief supervisors. Few were adequately trained.

Evaluations revealed that some maintenance departments even considered that work was adequately planned if there was even a slight pause between receipt of the work request and their response to it. For some mysterious reason, many found it necessary to manipulate statistics to convey the impression that a high percent of work was being planned. There were often faulty classification procedures in which it was difficult for anyone to determine what work was planned. Routine, repetitive PM services like weekly inspections or monthly lubrication tours were considered planned when, in fact, they were merely scheduled. Many large planning departments have failed to cut back their staffs when the use of the computer has made it possible to establish standard tasklists and material requirements for over 60% of work that can be planned. Generally, if the PM program is not generating deficiencies at least one week before work must be performed, there is little need for planners. They have not learned of requirements far enough in advance to plan them. Thus, the quality of planning is linked with the quality of PM inspection and testing.

Solution:
1. Examine the planning procedure.
2. Check to see if there is a criteria for determining which jobs will be planned and scheduled.

Establish a Responsive Work Order System

Evaluations uncovered not work order systems but rather one piece of paper—inadequate for detailed planning but overly complex for a simple request. No provision was made for the inevitable verbal orders. Efforts to suppress them resulted in no work order and no information about the job. The work was done and forgotten about. Standing work orders became a "burial ground" for costs and denied collecting repair history that should have been isolated unit-by-unit. Rarely was there an "engineering work order" to control non-maintenance project work like construction. Confusing attempts were made to try to use the maintenance work order to control this work when a contractor performed it.

Solution:
   Organize a task force of accounting, operations, material control, and maintenance to evaluate and correct the inadequacies that often emerge under scrutiny. Insist on a work order system that provides an element to control each type of work. Also, ensure that there is a link between the work order system and accounting to produce information on costs. Maintenance will welcome the evaluation because the work order system is its communication system. It affects everything they do.

Ensuring Accurate Field Labor Data Is Reported

Labor use reported by hourly personnel was revealed to be carelessly recorded with more attention paid to "filling up the 8 hours" than providing accurate, timely, and complete data. Some supervisors resorted to filling out craftsmen's time cards themselves to meet the accounting requirement. In most instances, the data was conjecture. Hourly personnel know they are the best source of information on what they actually did and should be given the opportunity to report it directly. Some crew members did not report accurately because the importance of reporting was not explained and they were suspicious of the use of reports. The end result was misleading information because of poorly reported field data. Most maintenance departments did poorly in reporting labor because of a lack of emphasis rather than a faulty reporting scheme.

Solution:
1. Ensure labor is reported initiated by the personnel who do the work and not by the supervisor who guesses at it. The supervisor should verify, not initiate the information.
2. Verify the accuracy, timeliness, and completeness of labor data by comparing it with actual work assignments.
3. Solicit the accounting viewpoint in finding ways to streamline labor reporting.

Remember that the only way maintenance can reduce the cost of doing its work is to improve its effectiveness in installing materials. This makes the control of labor a vital function and one that requires careful scrutiny.

Get the Best Material Control Support

Material control (inventory control and direct charge purchases) was usually well carried out when managed by material control professionals. When maintenance controlled parts inventory, there was little continuity between inventory control and direct charge purchasing. Usually, it split the material control function and created confusion. Poorly administered maintenance programs usually meant poorly administered inventory control programs when maintenance was responsible for the function. A major problem was the adequate identification of parts. Craft personnel identified parts poorly because of difficult procedures. As a result, supervisors were forced into this role. Parts identification tasks seriously reduced the time supervisors should have been in the field controlling work. In turn, this situation produced lower worker productivity.

Solution:
1. Check maintenance downtime against causes such as no materials available, wrong materials issued, or excessive time waiting for materials. These factors indicate the state of material control.
2. Make sure the material control function is not split with purchasing under accounting and stores under maintenance, for example. It seldom works effectively.
3. Watch stock room transactions to learn how well maintenance personnel actually go about identifying materials. During an evaluation, a surprising number simply came to the warehouse, gained entry, and "prowled" the aisles until they found what they needed—no records were made. They just took the parts. In one remarkable encounter, the maintenance man took two large pillow block bearings in case one got lost before he got around to installing it.
4. Check the supervisor's office. If it looks like a library of parts books, this clerical activity of parts identification may be the supervisor's principal job.
5. Check the "bootleg" storage areas if you wish to learn the awful truth of why material costs are so high.
6. Check the supervisor's desk drawers and wall lockers if you want to find out where the real stock room annexes are.
7. Do not threaten a forced clean up as too many usable items will end up in the waste dump rather than returned to the storeroom.

The resolution of material control problems begins with a physical check such as the one just outlined. However, the solution lies in a corrective effort between maintenance, accounting, purchasing, and the warehouse. Within maintenance, educate personnel on the material control program. Often they do not understand it.

Utilize Maintenance Information Systems Effectively

Information systems seldom covered decision-making information aspects adequately. These aspects include cost, repair history, backlog, labor utilization, the status of major jobs, and performance indices. There was an overabundance of administrative information on items such as parts cross-references, equipment lists, or absent personnel.

Thus, while everyone could easily look up parts, few could adequately manage maintenance because information systems emphasized administrative rather than management information. Systems provided by the corporate level, in an effort to achieve uniformity between several plants, did not work well. Some maintenance organizations purchased "stand alone" software packages only to find that they needed an integrated system but could not obtain local support to create necessary communications software. Some "package" programs were incompatible due to both language and logic considerations. Often commercial "packages" failed to link adequately with field labor and material data from time cards and stock issues or purchasing documents. Other "packages" were competent but overly complex and, as a result, personnel either misused them or ignored them. Often a new maintenance "package" proved incompatible with the existing inventory control program. But, rather than write communications software to link them, the companion "inventory control package" was purchased requiring the warehouse to replace its otherwise competent inventory control program. This proved expensive, time consuming, and disruptive. This not only illustrates the areas that maintenance impacts, but it suggests that information system implementors must know what they are doing. Generally, "package" vendors provided very little training and maintenance personnel had difficulty making the program perform up to its advertised capabilities. Most organizations were not satisfied with the timeliness, completeness, and accuracy of information.

Solution:
Determine whether maintenance has sufficient decision-making information to manage itself. Mandatory are:

1. *Labor utilization:* How effectively each supervisor controls his/her crew; whether absenteeism and overtime are controlled.
2. *Backlog:* The degree to which maintenance keeps up with the generation of work and the ability to adjust the work force size and craft composition as work load changes.
3. *Cost:*
   a. Data on equipment and components by which troublesome equipment can be identified.
   b. Functions like training, building repair, or snow removal.

    c. A summary showing the actual current month and year to date cost by cost center contrasted with the authorized budget.

4. *Status of major jobs:* Be able to isolate the cost and performance of selected major jobs from inception to completion and summarize this data on job completion.

5. *Repair history:* The chronological list of significant repairs made on key units of equipment identifying chronic, repetitive problems, failure trends, and the life span of critical components by which future replacement can be forecasted.

Once the right information has been obtained, ensure that it reaches decision makers and they take corrective actions.

In the selection or development of the maintenance information system, ensure that maintenance acts in concert with other departments in the selection and implementation of the system.

1. Operations must use the system to request work and check work status.
2. The warehouse must manage its inventory.
3. The purchasing agent must track order status.
4. Accounting must derive accurate field data for its general ledger programs such as payroll and accounts payable.

Make certain that the actual users, supervisors and craft personnel, are competent to use the system. Often, few have had any previous computer experience. If they are not trained, your considerable expenditure will produce no better results than the scraps of paper they used to write job notes on. If in-house Management Information System (MIS) support is used for implementation, insist on a strict time table. Assign the MIS personnel to the maintenance department until the job is complete so that their progress can be expedited.

Utilize Maintenance Engineering Effectively

Evaluations revealed that maintenance engineers were usually used on non-maintenance engineering projects like equipment installation or modification. Not enough attention was paid to functions which ensured equipment maintainability and reliability. A large percent of maintenance engineers were young and had degrees in a specific engineering discipline. They tended to gravitate toward what they understood best (project work) rather than bona fide maintenance engineering. Plants contributed to this problem by failing to adequately define maintenance engineering.

Solution:
1. Require that maintenance engineering focus on ensuring the reliability and maintainability of equipment.
2. Ensure that practical solutions like effective PM or repair standards are being instituted to reinforce reliability objectives.
3. Verify that a procedure exists to determine that newly installed equipment is able to be maintained. Are there maintenance instructions, spare parts lists, and wiring diagrams, for example?
4. Assure that maintenance engineers are not acting as project engineers and misused on non-maintenance activities such as construction and equipment installation.

Assess Maintenance Resource Use on Engineering Projects

Invariably, major engineering projects were found to be carried out with maintenance personnel rather than by a contractor or a segment of maintenance work force set aside for this work. Too often, this was not controlled and the maintenance program suffered from a misapplication of available labor. Usually these projects were well planned and organized because they were in the plant "spotlight." Lower cost projects were seldom funded correctly with questionable attempts made to expense the work against the maintenance budget rather than seek capital funds through proper channels. Projects tended to be executed on time, but often at the cost of diverting maintenance labor from necessary maintenance work. Most big projects had direct attention from management and monitoring by corporate personnel. These aspects, in part, explained their success.

Solution:
1. Curb the tendency to misuse maintenance resources on non-maintenance projects by developing and enforcing policies that preclude the unfair diversion of maintenance labor from its own program.

2. Check lower cost project funding practices to make sure the maintenance budget is not unfairly debited for a project that should have been capitalized.

Ensure Management and Staff Support for Maintenance

Evaluations often revealed that other departments showed an appreciation for the value of a good maintenance program, but lacked a specific means of providing support beyond saying "I'm for good maintenance."

Solution:
1. Specify instances in which operations can support maintenance effectively by making equipment available on time and reporting problems promptly.
2. Suggest positive solutions such as a weekly, joint operations maintenance scheduling meeting. Attend meetings to see how well they negotiate the work. Check schedule compliance.
3. Observe how well staff services such as warehousing are provided and, if inadequate, go to them and suggest how they might improve.

Attain Better Housekeeping

Evaluations found that housekeeping ranged from acceptable to excellent. There was a tendency in maintenance to allow discarded components to build up which created junk piles. There was poor follow-up in getting components into the rebuild pipeline, especially if done commercially.

Solution:
1. Extend housekeeping inspections into the well-known areas where maintenance has a tendency to collect junk in case they might be able to use it later.
2. Make sure the procedure for rebuilding components is well organized. Worn parts should be tagged and identified before they are sent to the warehouse for classification. Once returned from re-

build, parts should go back into the warehouse. When installed, track their performance to check the quality of rebuilds.

Motivate Maintenance Personnel

Evaluations suggested that hourly level personnel were in need of motivation while supervisors and planners were marginally better. Maintenance managers and superintendents appeared well motivated.

Solution:
1. Increase direct contact with supervisors and workers to uncover the reason for their poor motivation. You will create more opportunity to discuss common problems like working conditions and safety practices. Often, solutions will surface and morale will improve.
2. Be aware that motivation may be affected by administrative procedures such as the ease of parts identification or inadequate work assignment procedures. Check these matters and correct them.

Implement Team Organizations Correctly

Evaluations showed that those maintenance departments attempting a conversion from a traditional organization to a team had more difficulty than those starting with a team concept. A serious problem was the "about to be displaced supervisor" who saw elimination of his position, felt threatened, and resisted. Nonunion plants embraced the team concept more enthusiastically and younger, better educated workers created better teams. Invariably, the greatest difficulty experienced by all was traced to a poorly defined maintenance program. This oversight undermined the efforts of even the more sincere team efforts. In all instances observed, little progress was made until the maintenance program was properly defined.

Summary

In this era of intense competition, only the profitable plants will survive. Maintenance, which often represents over 30% of operating costs, is one of the few major costs that can be controlled. If costs are not controlled and reduced, they deny the profitability you seek. An ineffective maintenance program creates yet another unfavorable cost impact—equipment downtime. Downtime is three times as costly as the maintenance that could have avoided it. The lost product never reaches the marketplace and the downtime from poor maintenance diminishes profitability. An evaluation establishes the current maintenance performance level by identifying those activities needing improvement as well as those being performed well. The evaluation not only confirms performance but is the starting point for any improvement effort. Yet, the most important by-products of a well-conceived and effectively conducted evaluation are the education of plant personnel and their commitment to provide genuine support for improving maintenance. An evaluation that successfully points the way to improvement makes that improvement more attainable and is a matter of pride to maintenance. Thus, evaluations get the very people involved who can create the improvements to assure profitability. They will have been convinced that they can help because, as the manager, it was you who created the conditions for the success of maintenance. You confirmed in their minds that evaluations are not only the first step of improvement but an action to ensure their continuing improvement. Thus, your leadership in reviewing maintenance performance, creating an environment for their success, and providing a continuing impetus to evaluation and improvement represents the most positive steps you, as manager, can take to assure the maximum contribution of maintenance to the profitability of your plant.

Appendix A

Outline of Duties of Key Maintenance Personnel

Maintenance Superintendent: The superintendent is responsible for the entire maintenance function. He performs this through line supervisors such as general supervisor or first-line maintenance supervisor. He uses maintenance engineers in a staff capacity to ensure that the total maintenance program has continuity and that all new construction or installations can be supported by the maintenance program. Maintenance planners are used to develop plans for resource use on major work that must be scheduled to ensure the least interruption of operating schedules and best use of maintenance resources. Clerks are used to perform necessary administrative tasks related to pay, vacations, absenteeism, and so forth. In the instance of small organizations, the maintenance superintendent may control purchasing and stockroom functions if this is not done by accounting or some other plant department.

General Supervisor: The general supervisor controls two or more first-line maintenance supervisors. Normally, the general supervisor controls the planning activity which supports two or more supervisors. He divides his time about equally between direct field supervision and administrative duties such as planning, cost control, union business, and vacation planning.

Supervisor: The principal duty of the maintenance supervisor is the supervision of his crew and responsibility for carrying out the work load assigned to him. The supervisor generally carries out PM services, unscheduled and emergency work on his own initiative. Major planned and scheduled jobs appear on an approved schedule prepared by a supporting planner and jointly approved by the maintenance and operations superintendent. The supervisor must instruct and train crew members, arrange their vacation time, and discipline them as necessary. He prescribes work methods and procedures and ensures the timely completion of work by adequately supervising it. He is responsible for the conduct of PM services, the control of labor (including overtime), and the procurement of materials for unscheduled and emergency work.

Maintenance Engineer: The maintenance engineer is a staff person responsible to the superintendent for ensuring that equipment and facilities are properly installed, modified correctly, if needed, correctly maintained, and performing effectively. Specific responsibilities include assessment of newly installed equipment to ensure maintainability, parts, maintenance, instructions, prints, adequacy of design, and installation work. In addition, the engineer monitors ongoing work to ensure that sound, craftsmanship-like procedures are followed. He also observes the adequacy of stockage of correct parts in proper quantity. Further, he monitors the quality of parts used to ensure good performance and recommends action to correct inadequacies. He reviews repair history and costs to determine repair, rebuild, overhaul, or corrective maintenance needs. The maintenance engineer also reviews the conduct of PM services to ensure that they are conducted on time according to standards; he reviews equipment inspection deficiencies uncovered as a source of corrective maintenance. He also helps to develop standards for individual major jobs, procedures for performing work, cost levels, and the quantity of resources used. Periodically, he makes a cost-benefit review to determine make or buy actions. He also develops recommendations for training. As required, he prescribes methods for nondestructive testing (predictive maintenance). He constantly reviews the information system to ensure its adequacy and uses it to review the work load and backlog with the view of making work force level change recommendations.

Maintenance Planner: The maintenance planner provides direct support in planning major actions such as overhauls, rebuilds, or non-maintenance construction work prior to execution by maintenance supervisors and crews. These planning actions include job organization, labor estimating, identifying and obtaining materials, and coordinating work with production or engineering. Once work is scheduled, the planner allocates labor according to priority and monitors the control of the execution of planned and scheduled major jobs. In addition, the planner monitors conduct of the preventive maintenance equipment inspections. He also analyzes repair history to convert results into planned maintenance work. The planner should confine his activities to planned and scheduled maintenance and project work. Unscheduled and emergency work are controlled by maintenance supervisors.

Leadman: The crew leader is a regular maintenance craftsworker who, in consideration of an hourly wage increment and any special talents, can be used temporarily to help a supervisor coordinate elements of major crew jobs. The crew leader should not be considered as a supervisor since he cannot be expected to discipline fellow crew members.

Appendix B

Typical Maintenance Objective

THE MAINTENANCE OBJECTIVE

The maintenance objective, an extension of the production strategy, links the task of maintenance directly to that strategy. The primary objective of maintenance is the repair and upkeep of production equipment to ensure that it is kept in a safe, effective, as designed, operating condition so that production targets can be met on time and at least cost. A secondary objective of maintenance is to perform approved, properly engineered and correctly funded non-maintenance work (such as construction and equipment installation) only to the extent that such work does not reduce the capability for carrying out the maintenance program. In addition, maintenance will operate support facilities (such as power or steam-generation), but will ensure that necessary resources are allocated within its authorized work force and are properly budgeted. As appropriate, maintenance will also monitor the satisfactory performance of maintenance contract services.

There are several key phrases used in the maintenance objective, each with a specific intent. If you so intend, a primary as well as a secondary objective might be provided to establish clear precedence. That is, production equipment maintenance is first while project work follows, resources permitting. Further, the use of phrases such as "as designed" means that equipment modification is excluded from the primary maintenance task (it is not maintenance). This implies that you might wish all non-maintenance work to be approved, properly engineered, and correctly funded. You may also intend that such work may only be done if it does not reduce the capability of maintenance to meet its primary objective. If maintenance is to perform operating functions such as power-generation you should ensure that they are properly staffed and budgeted to do so. With an objective such as that cited, there should be little question about maintenance priorities and limits as to what and how much it can do. Based on such the assigned objective maintenance knows its responsibilities exactly and may organize properly to carry them out. Its customers are also aware of maintenance limitations and will request support accordingly.

The objective also helps to ensure that the intended maintenance program can be carried out effectively. The absence of a clear objective can yield unfortunate consequences. For example, in a large paper mill complex, the 300-person maintenance work force was having difficulty carrying out the maintenance program. During shutdowns it was necessary for them to be augmented by contractors to catch up. An investigation showed that up to 40% of work requests made by production were not maintenance. Rather, they were equipment changes, modification, installations, and so forth. As a result, nearly 35% of maintenance manpower was used on this work detracting from manpower available for basic maintenance. Once the plant manager saw these circumstances, he clarified the maintenance objective and educated personnel in the proper use of the maintenance work force. The objective helped to carry out this task successfully.

Appendix C

Illustrative Policy Guidelines for Maintenance

MANAGEMENT POLICY GUIDELINES

Only you, as the manager, can establish policy. Your maintenance policies ensure that important aspects of the plant production strategy, as they apply to maintenance, are understood and followed uniformly plant-wide. Your policies provide the basis on which day-to-day maintenance procedures are built. Thus, for example, if you place sufficient importance on preventive maintenance, and spell out roles to be played by maintenance in creating the program, and of production in complying with it, the end result will be a more successful program. A typical policy on preventive maintenance might be:

1. Maintenance will conduct a "detection oriented" Preventive Maintenance (PM) program. The program will include equipment inspection and nondestructive testing (predictive maintenance) to help avoid premature failure; lubrication, servicing, cleaning, adjusting, and minor component replacement to extend equipment life.
2. PM will take precedence over every aspect of maintenance except bona fide emergency work.
3. Production will perform PM-related tasks such as cleaning and adjustment and ensure that all services due are carried out on time.
4. Compliance with the PM schedule will be measured and management informed of performance improvements as a result of PM services performed.

Note the policy states that the purpose of PM as well as the requirement of production is to support the program. Such a policy demonstrates your interest in assessing compliance and measuring benefits achieved.

One organization learned that, without a management policy on PM, maintenance had difficulty performing its program while operations seemed to consider PM a discretionary option. At a large distribution warehouse, a fleet of 80 lift trucks was in use. The PM program for the lift trucks was carefully designed and judged, on paper, to be first-rate. However, it was ineffective. An investigation revealed that production seldom brought units to the garage for PM services. Rather, they required maintenance to come after the units, often having to track them down, sometimes unsuccessfully. Realizing that maintenance had been asked to accomplish the impossible, the manager directed that operations be responsible for verifying that scheduled PM services were carried out on time, and he be informed of schedule compliance. Within a month, the lift truck PM program was significantly better.

Your policy guidelines on maintenance should span the entire maintenance program. They must not be limited to a few key elements. Policies will be the basis of maintenance field procedures and instructions. Therefore, they should be flexible enough to permit practical interpretation. Following is a list of typical policy guidelines.

1. On department relationships
   a. Production will be responsible for the effective utilization of maintenance services.
   b. Maintenance will be responsible for developing a pertinent maintenance program, educating personnel on its elements, and carrying it out diligently while making effective utilization of its resources and ensuring quality work is performed.
   c. Each department manager will ensure compliance with the policies covering the conduct of maintenance.
2. On the maintenance program
   a. Maintenance will publish work load definitions and appropriate terminology to ensure understanding and proper usage.
   b. Maintenance will conduct a "detection oriented" Preventive Maintenance (PM) program. The program will include equip-

ment inspection and nondestructive testing to help avoid premature failure; lubrication, servicing, cleaning, adjusting, and minor component replacement to extend equipment life. PM will take precedence over every aspect of maintenance except bona fide emergency work.

c. A work order system will be used to request and control work.

d. Planning and scheduling will be applied to comprehensive jobs (such as overhauls and major component replacements) to ensure that work is completed productively and expeditiously.

e. Maintenance will publish a priority-setting procedure which allows other departments to communicate the seriousness of work and maintenance to effectively allocate its resources. The procedure will facilitate the assignment of the relative importance of jobs and the time frame within which the jobs should be completed.

f. Maintenance will develop and use information concerning the utilization of labor, the status of work, backlog, costs, and repair history to ensure effective control of its activities and related economic decisions such as equipment replacement. Minimum necessary administrative information will be developed and used. Performance indices will be used to evaluate short-term accomplishments and long-term trends.

3. On the control of labor

a. The maintenance work load will be measured on a regular basis to help determine the proper size and craft composition of the work force.

b. The productivity of maintenance will be measured on a regular, continuing basis to monitor progress in improving the control of labor.

4. On maintenance engineering

a. Maintenance engineering will be emphasized to ensure the maintainability and reliability of equipment.

b. Current technology will be utilized to facilitate effective maintenance.

c. No equipment will be modified without the concurrence of maintenance engineering.

d. All new equipment installations will be reviewed by maintenance engineering to judge their subsequent maintainability.

5. On material control
   a. Procedures for obtaining stock and purchased materials or services will be strictly adhered to.
   b. Parts will not be removed from any unit of equipment and used to restore another unit to operating condition without explicit authorization from the maintenance superintendent.
6. On non-maintenance work
   a. Engineering, operations and maintenance are jointly responsible for ensuring that all non-maintenance projects (construction, modification, equipment installation, and so forth) are necessary, feasible, properly engineered, and correctly funded before work commences.
   b. Maintenance is authorized to perform project work such as construction, modification, equipment installation, and relocation only when the maintenance work load permits. Otherwise, contractor support will be obtained subject to the current labor agreement.
   c. Equipment modifications will be reviewed to determine their necessity, feasibility, and correct funding prior to the work being assigned to maintenance. All such work will be reviewed by maintenance engineering before work commences.

Appendix D

Maintenance Work Load Definition

MAINTENANCE WORK LOAD DEFINITION

Maintenance Work   The repair and upkeep of existing equipment, facilities, buildings, or areas in accordance with current design specifications to keep them in a safe, effective condition while meeting their intended purposes.

Scheduled Maintenance   Extensive major repairs such as rebuilds, overhauls, or major component replacements requiring advanced planning, lead time to assemble materials, scheduling of equipment shutdown to ensure availability of maintenance resources including labor, materials, tools, and shop facilities.

Preventive Maintenance   Any action which can avoid premature failure and extend the life of the equipment. It includes equipment inspection and testing to avoid premature failure and lubrication, cleaning, adjusting, and minor component replacement to extend equipment life.

Emergency Repairs   Immediate repairs needed as a result of failure or stoppage of critical equipment during a scheduled operating period. Imminent danger to personnel and extensive further equipment damage as well as substantial production loss will result if equipment is not repaired immediately. Scheduled work must be interrupted and overtime, if needed, would be authorized in order to perform emergency repairs.

Unscheduled Repairs   Unscheduled nonemergency work of short duration often called "running repairs." Work that can be accomplished within approximately 1 week with little danger of equipment deterioration in the interim. Repairs are usually performed by one person, often in 2 hours or less, and in about 40% of instances parts are not required.

Routine Maintenance (repetitive work)   Janitorial work, building and grounds work, training, safety meetings, shop clean-up, and highly repetitive work such as tool sharpening.

NON-MAINTENANCE WORK LOAD

Project Work   Construction, installation, relocation, or modification of equipment, buildings, facilities, or utilities. Usually capitalized.

Construction   The creation of a new facility or the changing of the configuration or capacity of a building, facility, or utility.

Installation   The installation of new equipment.

Modification   The major changing of an existing unit of equipment or a facility from original design specifications.

Relocation   Repositioning major equipment to perform the same function in a new location.

Appendix E

Maintenance Terminology

Adjustments   Minor tune-up actions requiring hand tools, no parts, and less than one half hour.

Administrative Information   Information used to communicate within maintenance and operate the maintenance information system.

Area Maintenance   A type of maintenance in which one supervisor is responsible for all maintenance within a reasonably sized geographical area.

Backlog   The total number of estimated man-hours, by craft, required to complete all identified, but incomplete planned and scheduled work.

Capital Funded   Non-maintenance work authorized by a capital fund authorization.

Capitalized   Funding for work which expands plant operating capacity, gains economic advantage, replaces worn, damaged or obsolete equipment, satisfies a safety requirement, or meets a basic need.

Category   The types of work which make up the work load performed by maintenance (preventive maintenance and emergency repairs).

Component   A sub-element of a unit of equipment (the belt of a conveyor, the motor of a crusher, the engine of a truck).

Concept   The means by which a major program, such as maintenance, is carried out in relationship to its objective and the other programs which it supports (operations) or on which it depends (purchasing).

Coordination   Daily adjustment of maintenance actions to achieve the best short-term use of resources or to accommodate changes in operation needs.

Cost-Center   A department or area in which equipment operates or in which functions are carried out.

Decision-making Information   Information necessary to control day-to-day maintenance and determine current and long-term cost and performance trends for management decisions.

Deferred Maintenance   Maintenance which can be postponed to some future date without further deterioration of equipment.

Downtime   A time period during which equipment cannot be operated for its intended purpose.

Engineering Work Order   A control document authorizing use of the maintenance work force or a contractor for engineering project work such as construction.

Equipment and Function File   A computer file which lists all equipment by type, with its components, plus functions carried out within a cost-center.

Expensed   Maintenance work charged to the operating budget.

Failure Coding   An indexing of the causes of equipment failure on which corrective actions can be based.

Forecasting   A projection of anticipated major tasks that are predictable based on historical data.

Function   An activity carried out on a unit or performed within a cost-center (250 hour service on haulage truck number 110, power-sweeping in department 06).

Functional Maintenance   Maintenance in which the supervisor is responsible for conducting a specific function like pump maintenance for the entire plant.

Inspection   The checking of equipment to determine repair needs and their urgency.

Level-of-Service   The degree of maintenance performed to meet desired levels of equipment performance. A high level ensures little chance of failure while a low level meets minimum requirements risking breakdowns on less critical equipment.

Maintenance Engineering   The use of engineering techniques to ensure equipment reliability and maintainability.

Maintenance Information System   A means by which field data is converted into information so that maintenance can determine work needed, control the work, and measure the effectiveness of the work done.

Maintenance Work Order (MWO)   A formal document for controlling planned and scheduled work.

Maintenance Work Order System   A means of requesting maintenance service, planning, scheduling, controlling work, and focusing field data to create information.

Maintenance Work Request (MWR)   An informal document for requesting unscheduled or emergency work.

Major Repairs   Extensive, non-routine, scheduled repairs requiring deliberate shutdown of equipment, the use of a repair crew possibly covering several elapsed shifts, significant materials, rigging, and, if needed, the use of lifting equipment.

Minor Repairs   Repairs usually performed by one worker using hand tools, few parts, and usually completed in less than one half shift.

Nondestructive Testing   See Predictive Maintenance.

Objective   The principal purpose for the existence of each line department (like maintenance) or staff department (like purchasing) and the roles that they must play to assure that the plant production strategy is achieved.

Overhauls   The inspection, teardown, and repair of a total unit of equipment to restore it to an effective operating condition in accordance with current design specifications.

Performance Indices   Ratios which convey short-term accomplishments and long-term trends against desired standards.

Periodic Maintenance   Maintenance actions carried out at regular intervals.

Planning   Determination of resources needed and the development of anticipated actions necessary to perform a scheduled major job.

Policies   Management guidelines for the development of field procedures to assure achievement of plant profitability.

Predictive Maintenance   Nondestructive testing techniques to predict wear rate, determine state of deterioration, monitor condition, or predict failure.

Principles   Logic, common sense, proven procedures, or essential rules on which plant operation must be based.

Priority   The relative importance of a single job in relationship to other jobs, operational needs, safety, equipment condition, and the time within which the job should be done.

Procedure   The day-to-day method for carrying out elements of the maintenance program such as assignment and control of work. Field procedures should be built on policies.

Production Strategy   The plan for achieving plant profitability.

Productivity   The percentage of time that maintenance personnel are at the work-site, with their tools, performing productive work during a scheduled working period.

Project Work   Actions such as construction, equipment modification, installation, or relocation to gain economic advantage, replace worn, damaged or obsolete equipment, satisfy a safety requirement, attain additional operating capacity, or meet a basic need. Usually capital funded.

Purchase Order   The authorized document for obtaining direct charge materials or services from vendors or contractors.

Quality Standard   A standardized procedure for accomplishing a major maintenance task.

Quantity Standard   The standard resources required to meet the prescribed
quality standard.

Rebuild   The repair of a component to restore it to serviceable condition in accordance with current design specifications.

Relocate   Move fixed equipment to a different location.

Repair History   A record of significant repairs made on key equipment used to
spot chronic, repetitive problems, failure patterns, and component lifespan
which, in turn, identifies corrective actions and helps forecast component replacements.

Repetitive Maintenance   Maintenance jobs which have a known labor and material content and occur at a regular interval.

Reposition   Move mobile equipment to a new working location.

Routine Maintenance   Maintenance or services performed consistently in the
same manner.

Schedule Compliance   The effectiveness with which an approved schedule was
carried out.

Scheduling   Determination of the best time to perform a planned maintenance
job to appreciate operational needs for equipment or facilities and the best
use of maintenance resources.

Specifications   Technical definition of equipment configuration or performance
requirements to meet intended utilization of equipment or materials.

Standing Operating Procedures (S.O.P.)   A written procedure used to ensure
reasonable uniformity each time a significant task is performed.

Standing Work Order   A reference number used to identify a routine, repetitive
action.

Stock Issue Card   The authorized document for making stock material with-
drawals.

Time Card   The authorized document for reporting the use of labor.

Type   All equipment of the same kind (conveyors, pumps, haulage trucks, loaders).

Unit   One unit of equipment of a specific type (conveyor 006).

Utilization   The percentage of time that a maintenance crew is available to perform productive work during a scheduled working period.

Verbal Orders   A means of assigning emergency work when reaction time does
not permit preparation of a work order document.

Work force   The personnel who carry out the maintenance work load.

Work load   The essential work to be performed by maintenance and the converversion of this data into a work force of the proper size and craft composition to ensure that the program is carried out effectively.

Work Order System   A communications system by which maintenance work is requested, classified, planned, scheduled, assigned, and controlled.

Work Sampling   The statistical measure of the utilization of labor to determine productivity.

Appendix F

Typical Maintenance Self-Evaluation Questionnaire

PREVENTIVE MAINTENANCE (PM)

The preventive maintenance (PM) program should successfully extend equipment life and avoid premature failures through timely inspection, testing, lubrication, cleaning, adjustment, and minor component replacements. As a result, there should be fewer emergency jobs and more work should be able to be planned. As the planned work is performed, maintenance personnel will work more productively and the results will have lasting quality.

Rate the organization, execution, and effectiveness of preventive maintenance by comparing the following standards with the performance of your maintenance department based on your personal knowledge. Mark your ratings on an accompanying answer sheet. Rate 1 to 10 (highest) if you have personal knowledge on maintenance performance. Mark "X" if you do not know. If a (0) appears on your answer sheet, do not respond since your job does not require that you evaluate this standard.

1. There is an effective overall PM program.
2. Plant management understands and strongly supports PM.
3. The PM program is oriented toward uncovering deficiencies before equipment fails.
4. The PM program emphasizes safety.
5. The PM program emphasizes the preservation of assets.

6. There is evidence that the PM program has reduced the amount of emergency work.
7. As the result of the PM program, more work is being planned.
8. The manpower required for each PM service and for the entire PM program is known.
9. PM services are verified for quality and adherence to the schedule.
10. New equipment is added to the PM program promptly.
11. The PM program is reviewed periodically and updated to reflect changing conditions.
12. Maintenance personnel conduct PM services effectively.
13. Maintenance supervisors ensure that PM services are performed effectively and on time.
14. The PM program has been explained to operating personnel to enable them to cooperate and use its services effectively.
15. Operating personnel cooperate with the PM program and perform simple, routine PM related tasks to help ensure dependable operation of equipment.
16. Appropriate nondestructive testing techniques such as vibration-analysis and infrared testing have been identified and, as required, integrated into the PM program.
17. Each PM service has a standardized checklist which explains how and when the service is to be performed.
18. Each PM action is identified by a code or number to aid in scheduling, control, and reporting.
19. Extensive repair actions during the conduct of PM services, especially inspections, are avoided.
20. The timing of PM services is carefully regulated according to fixed time intervals (fixed equipment) or accumulated operating hours and miles (mobile equipment).
21. PM services for individual units of fixed equipment are linked together in routes to avoid unnecessary travel time or backtracking.
22. PM services for mobile equipment are scheduled in advance to avoid unnecessary interruption of operations.
23. Individual operators and maintenance workers cooperate in the conduct of PM services.

Appendix G
Action Group and Decision Group

The successful implementation of improvements in maintenance will require changes in behavior of not only key personnel but workers as well. Since change is difficult for many adults, you may expect resistance to change in implementing improvements. Among the most effective means of reducing resistance to change is education and participation in the change process. Typically, if a person understands why a change is necessary, he will be more willing to consider how such a change can be brought about. Then with his direct participation, he can be in a position to influence exactly how the change is to be made. Often, if it affects him directly, his participation will remove any misunderstanding and potential threat to him or his feeling of security. After a decision has been made on what the change should be, it is advisable to test the changes in the working environment. Thus, there is a further opportunity to modify the change to better suit the operating circumstances. But, more importantly, the participants come to realize that these modifications, which they also suggest, increase the control that they have over their own destiny. Thus, with education and participation, resistance is soon displaced with acceptance of changes and positive, willing support. The end result is improved performance through the successful implementation of changes suggested by the evaluation.

TECHNIQUE

Advisory Groups

The use of two advisory groups has proven to be a very effective way to bring about the required changes while educating them and providing opportunity for participation. An action group and a decision group are typical of the way personnel can be organized to ensure active involvement in the implementation of changes in the maintenance organization and its program. Note the following definitions.

Action Group: A group of middle management and working personnel who must ensure that the maintenance program gets carried out and who recommend practical ways to implement maintenance program or organizational changes. Typically, the action group would consist of a mechanic, an electrician, an operator, a maintenance supervisor, and an operating foreman.

Decision Group: A group of management personnel who will approve or modify recommendations made by the action group prior to their implementation. Typically, the decision group would consist of the plant manager, the maintenance superintendent, the operating superintendent, and staff personnel like the personnel manager or accounting manager.

Coordinator: The coordinator is the link between the action group and the decision group. He/she would schedule meetings, prepare agendas, and document recommendations and actions approved by the decision group. The coordinator may be the maintenance superintendent, the planner, or a foreman not in the action group.

Secretary: Any person who can provide word-processing support in the preparation of agendas or summaries and recommendations. Files would be created since the recommendations, as subsequently tested, would become the future maintenance manual.

TYPICAL TASKS

Among the tasks that might be performed by the Action Group are:

1. Recommend training for craft personnel.
2. Help define maintenance terms.

3. Review policies for practical interpretation.
4. Look over the PM program for consistency.
5. Examine the method of identifying stock materials.
6. Review the criteria for work to be planned.
7. Review the method for recording modifications.

Meetings

The action group would meet weekly (during lunch) for no more than 45 minutes. Each meeting would have a specific agenda dealing with a single topic. Written material corresponding to the topic would be provided. Each meeting on a new topic would define the problem clearly before the action group develops possible recommendations. Some topics might be under discussion for several meetings. Recommendations need not be written, but they must deal with the specific topic. At subsequent meetings, verbal recommendations will be made and recorded. Summaries of recommendations will be prepared for the record and to brief the decision group. Every 4 weeks the action group would meet with the decision group to present its recommendations. Resulting decision group decisions, if not provided during the meeting, would be communicated to the action group subsequent to the monthly meetings.

Membership

Each designated member of the action group would have a back up who will attend if the designated member cannot attend. Every 4 weeks at the meeting with the decision group, the designated members as well as the back up will attend. At the conclusion of this meeting, the designated members will drop out of the group to be replaced by their back up. As this happens, new back up members will be appointed. The same personnel may be in the action group, but membership for any two consecutive one month periods should be avoided if possible.

Records

A notebook should be provided to each member of the action group to keep agendas, explanatory materials, and summaries. The notebook will be brought to each meeting. It should be kept where either the designated or back up member can have access to it.

Outside Input

Action group members are encouraged to discuss topics with anyone who can provide constructive comments.

Pilot Testing

Once a course of action has been agreed upon, the activity should be tested in a pilot area. Action group members would observe results and as necessary modify the procedure being tested (with the advice of those operating the test area). Further testing would continue until the desired performance level is achieved. Thus, the final result represents the full participation of all personnel involved while providing them with an education on the changes themselves and why they are necessary.

Appendix H

Measuring Maintenance Productivity

The maintenance department that checks productivity regularly is rare—perhaps even nonexistent. Despite this, productivity measurement remains the most effective way of verifying the quality of labor control. Numerous reason are offered for not checking productivity, among them:

1. Measurements will be misunderstood.
2. The work force will suspect layoffs.
3. Personnel will feel threatened.
4. Measurements are not accurate.
5. We do not trust industrial engineers.
6. We do not have time to check productivity.

Managers confronted with such resistance soon conclude that it is counterproductive to ask maintenance to verify its quality of labor control. In such an environment, even skilled industrial engineers are reluctant to conduct measurements despite strong management sanction. The result is that nothing happens. Productivity, if poor, remains that way. Worse yet, productivity that could easily be improved is not, current levels are not known, and the factors that inhibit productivity are not identified.

WHY NOT LET HOURLY EMPLOYEES MEASURE PRODUCTIVITY?

How can it be done? Briefly, if a measure is made of the factors that inhibit productivity, it is possible to deduce the level of productivity. Moreover, the factors that cause a loss of productivity can be clearly identified so that corrective actions can be initiated. It can be successfully argued that the best source of information on work performed by maintenance is the maintenance worker himself. Therefore, by extension, workers are in a position to comment successfully why they have difficulty getting their work done. Additionally, the trained industrial engineer can effectively measure only about five productivity related factors: waiting, travel, clerical, idle, and working. By contrast, the involved maintenance worker can, if properly trained and oriented, successfully measure numerous elements on time spent or time lost, typically:

- Identifying parts.
- Obtaining parts.
- Waiting because there is no work available.
- Due to incorrect instructions.
- Tools are not available, broken or not enough.
- Drawings are not up-to-date or cannot be found.
- Waiting for another craft.
- Waiting for equipment to be shut down.
- Waiting for lifting equipment.
- Excessive clerical tasks.
- Using archaic repair techniques.
- More or better training is needed.
- Union or labor related problems.

In this age of involvement, the best weapon we have to reduce resistance is participation—why not use it? If used successfully, most of the reasons offered by maintenance for not measuring productivity soon disappear.

TECHNIQUE

Random sampling is the key. Not every maintenance worker need be involved in the measurements. However, all workers should be aware of how the measurements are to be made, why, and what they reveal.

To start, there must be total agreement that the improvement of worker productivity is a common objective of management to utilize labor as effectively as possible, and the work force to be able to carry out work efficiently and with the satisfaction of knowing that quality work is being done. Following are the steps involved:

1. Step One
   a. Meet with work force representatives (or union officials) and maintenance supervision.
   b. Openly discuss the reasons, on both sides, for improving productivity.
   c. State the policy of improving productivity to reduce downtime, not reduce the work force.
   d. Assure personnel that the work force will not be reduced.
   e. Explain what will be done with the results of productivity measurements.
   f. Describe how data on measurements will be shared.
   g. Explain how joint decisions will be made in determining how to correct and reduce the factors inhibiting productivity.
2. Step Two
   a. Identify the frequency of measurements (6 months).
   b. Establish the periods when measurements will be made (weeks 21 and 22).
   c. Explain how supervisors will randomly select 10% of their crews to participate on alternate days during the measurement period.
   d. Demonstrate how each worker will select the longest job performed on the day he participates and record the time spent on each of the factors affecting productivity as well as the time spent working.
   e. Explain that no worker will be identified by name.
   f. Show how the data will be summarized to reveal the factors that must be improved, the degree, and the priority of improvement.
   g. Reach agreement on how management and worker groups will analyze the data and establish a plan for correcting and improving the problems identified.
   h. Determine how results will be shared plant-wide and comments welcomed.
   i. Establish the period for the next measurement.

SOME CONCLUSIONS

Productivity will not improve itself. Human beings must dedicate themselves to productivity's improvement. Both management and the work force suffer from poor productivity. Sufficient leadership must be exercised to call attention to this need and then to do something about it. The technique described is a possible starting point in improving productivity. It bridges the communications gap that has, for so long, blocked necessary action in measuring productivity and getting improvement started.

Index